Strategy in Emerging Markets

Studies in Global Competition
Edited by John Cantwell, The University of Reading, UK and David Mowery, University of California, Berkeley, USA

Volume 1
Japanese Firms in Europe
Edited by Frédérique Sachwald

Volume 2
Technological Innovation, Multinational Corporations and New International Competitiveness: The Case of Intermediate Countries
Edited by José Molero

Volume 3
Global Competition and the Labour Market
By Nigel L. Driffield

Volume 4
The Source of Capital Goods Innovation: The Role of User Firms in Japan and Korea
By Kong-Rae Lee

Volume 5
Climates of Competition
By Maria Bengtsson

Volume 6
Multinational Enterprises and Technological Spillovers
By Tommaso Perez

Volume 7
Governance of International Strategic Alliances: Technology and Transaction Costs
By Joanne E. Oxley

Volume 8
Strategy in Emerging Markets: Telecommunications Establishments in Europe
By Anders Pehrsson

Volume 9
Going Multinational: The Korean Experience of Direct Investment
Edited by Frédérique Sachwald

Strategy in Emerging Markets

Telecommunications Establishments in Europe

Anders Pehrsson

Växjö University
Sweden

London and New York

First published 2001
by Routledge
11 New Fetter Lane, London EC4P 4EE

Simultaneously published in the USA and Canada
by Routledge
29 West 35th Street, New York, NY 10001

Routledge is an imprint of the Taylor & Francis Group

Typeset by Expo Holdings, Malaysia
Printed and bound in Great Britain by MPG Books Ltd, Bodmin

British Library Cataloguing in Publication Data
A catalogue record for this book is available from the British Library

ISBN: 0–415–27052–9

CONTENTS

FIGURES

TABLES

PREFACE

This book relies on original strategy research in terms of development of concepts and models relevant to strategy in emerging markets in general, and of empirical studies concerning telecommunications operators in Europe. A cross-sectional approach was applied in the study of operators competing in the UK and Sweden, while the case study approach was used for penetration of strategies of specific operators. This book is an extension of my previous work on international strategies, notably developments concerning systems suppliers within telecommunications (Pehrsson, 1996).

Although the results and shortcomings of this book are entirely mine, I am grateful for discussions with colleagues. Valuable comments have been given by Professor John Cantwell, University of Reading, Berkshire, UK, Dr Dimitrios Ioannidis, Stockholm School of Economics, Sweden; and Professor Peter Lorange, The International Institute for Management Development, Lausanne, Switzerland.

Associate Professor Sikander Khan has shared the responsibility for our research programme, International Business Strategy. Moreover, he and Dr Lars Ehrengren, Mrs Eva Carnestedt, Mrs Eva Erichsen, Mrs Ann Lindgren and Mrs Linnea Shore in the School of Business, Stockholm University, have been helpful in organizing courses relevant to the book subject (that is, courses on international marketing, international business strategy, strategy research and executive courses). Dr Bengt Gustavsson and Magnus Gustavsson have also taught these courses and have thus applied earlier versions of the models presented in this book. Thanks are also due to all PhD students, notably Henrik Friman and Cranmer Rutihinda, and MBA students who provided valuable insights in discussions.

I would also like to thank here Michael Bryan-Brown, Manager of Regulatory Affairs of COLT Telecommunications in London;

Andrew Law, Head of Research and Analysis of Cable & Wireless Communications in London; and Pelle Hjortblad, Market Director of Tele2 in Stockholm. These persons invited me for interviews and supplied written information in the case studies.

A preparatory part of the project received financial support from the Nicolin Foundation "CN 70" of the Swedish National Committee of the International Chamber of Commerce, while the Tore Browaldh Foundation of Handelsbanken in Sweden, and the Swedish Transport & Communications Research Board supported the main project. Mrs Margareta Jacobson of the School of Business assisted in the accounting of the latter support.

Finally, I would like to dedicate this book to my wife Ann-Helene who has been supportive throughout the research process, and to our children Tobias, Terese, and Andreas.

GLOSSARY

AMPS: Advanced Mobile Phone System. The original American standard for analog systems.

Cellular mobile telephony: Mobile telephony system consisting of radio base stations linked together by telephone exchanges. Each station covers a geographical area, a cell. When a subscriber moves within an area covered by a system the call is transmitted from cell to cell.

GSM: Global System for Mobile Communication. It was originally developed as a European standard for mobile telephony but has become widely used throughout the world.

ISDN: Integrated Services Digital Network. A network in which information types such as voice, data and images is conveyed simultaneously to a subscriber through a common local line.

LAN: Local Area Network. A limited data network covering a limited geographical area such as a building.

NMT: Nordic Mobile Telephony. The Nordic standard for analog mobile telephony which was established in the early 1980s in Sweden, Finland, Norway and Denmark. Later, it was also installed in other countries.

PCS: Personal Communications Services. Collective term for American mobile telephony services in the 1900 MHz frequency band.

Radio in the local loop: Network solutions in which radio technology is used to link the subscriber to the wire-line telecommunications network.

ONE

Introduction

Markets that previously have been unreachable for companies other than monopolies or other protected firms are frequently being opened up. This applies to entire industries that become reachable through deregulation initiatives or otherwise, and it is also valid for geographical regions, such as certain countries in East Europe and Asia. This allows for the greater participation of both domestic and foreign firms, and, as a result, it allows for more intense competition. Some of the firms in question have been established previously, while some have been established in order to exploit opportunities in the emerging markets.

Extensive new establishments generally change the character of a specific industry, no matter how newcomers comprehend the situation, what competencies they possess, or how they enter the industry. More intense competition in an emerging market manifests itself in different ways. That is, different strategies may be developed in the market, and, frequently, the technological edge or other competitive edges may be decisive in the survival of the companies involved. All of these changing premises make the formulation of strategy in emerging markets a difficult endeavor.

But before we are able to shape detailed prescriptive models for the formulation of strategies in emerging markets, we need to extend our understanding of strategy development based on descriptions of realized establishment processes and initial strategies in the markets.

This book presents comprehensive descriptions of the industry of telecommunications operators in the UK and Sweden and provides detailed descriptions of strategy development of single operators competing in the two markets. These descriptions make it possible to detect patterns of strategy in the emerging markets in question, specifically establishments and initial business strategies in the markets. Furthermore, delineation of corporate processes of restructuring and integration consistent with strategies are facilitated by the empirical descriptions.

The industry in focus is subject to comprehensive international deregulation processes, including deregulation within the European Union, in Japan and in the USA. As early steps in the coming worldwide process were taken in the UK around 1980, it is of great interest to study the development of strategies of operators in this market. Deregulation also started early in Sweden, and it is therefore interesting to compare these two markets.

The telecommunications industry is one of the largest industries in the world. According to the International Telecommunications Union, in 1994, the European Union member states generated telecommunications revenue of approximately GBP 95 billion. The five largest countries by revenue were Germany, the UK, France, Italy, and Spain.

Thus, the telecommunications industry is subject to far-reaching changes that strongly affect market conditions for participating companies. Worldwide deregulation processes pertaining to network operators, intensive technology development, and growing demand create emerging markets within the industry. The emerging markets will cause all industry actors, not just operators, to reconsider their strategies and to take into consideration the effects of changing markets.

Generally speaking, there are five homogeneous types of actor in the telecommunications industry (Figure 1.1). Although the figure distinguishes between the five types, a single company may, in reality, take on more than one role, displaying a number of relationships to other actors in the market in question.

Equipment suppliers provide components and equipment for telephones, exchanges, base stations in mobile telephony systems, and networks. Some equipment constitutes the basis of systems and networks, and other types of equipment, such as mobile telephones, are bought by end-users from a retailer.

Operators of networks for stationary or mobile telephony offer telecommunications services and subscriptions to end-users (that

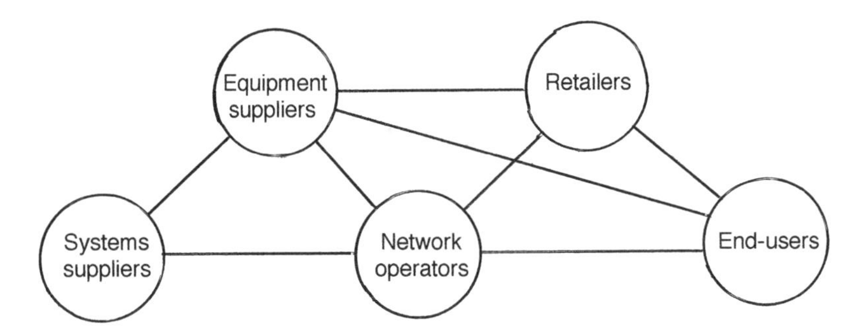

Figure 1.1 Types of actor in the telecommunications industry.

is, to individual consumers and companies, either directly or through retailers or other distributors). Distributors are frequently local companies comprising specialists and retail chains. A network operator normally has a close relationship with the particular systems supplier who has delivered the network in question. Thus, a pure systems supplier does not have direct contact with retailers or end-users.

This chapter continues with a discussion of the research problem and the research design, including a summary of the framework used in the book. Finally, a brief review of relevant changes in the telecommunications industry is presented. These changes, notably deregulation processes and technology development, affect industry actors to a large extent. The dynamic character of the changes is, however, manifested by the fact that I do not treat the processes of deregulations and technology development as static inputs into the formulation and implementation of strategy in emerging markets. Rather, I try to pay attention to the notion that industry actors themselves are more or less influential in shaping the processes.

CONTENT AND OUTLINE OF THIS BOOK

The direction and design of the research presented in this book are rooted in our need for more knowledge about strategy in emerging markets. Consequently, this section starts with a discussion of our need for knowledge and proceeds with a definition of the subject studied and an outline of the book.

What do we need to know?

Deregulation processes pertaining to industries have been studied in different ways. In fact, O'Reilly (1995) suggests that deregulation has been so extensive that it is possible to discern recurring changes irrespective of the industry in which the deregulation has occurred. Applying Porter's (1980) "driving forces of competition," O'Reilly observed deregulations that have taken place within the airline industry and categorised the process into five life-cycle phases (regulated strategic torpor, pre- deregulation jockeying, advent deregulation, shake-out, relative competitive stability). Throughout these phases, the intensity of competition varies due to the threats of potential entrants and subsititute technology, and the bargaining power of buyers and suppliers.

I think, however, that this life-cycle perspective represents a deterministic way of viewing strategy development in emerging markets. It generates universal models that suggest a strategic behavior independent of particular industry features and firm characteristics. All deregulating industries in all markets do not necessarily follow the same pattern. Furthermore, it is very doubtful whether the content of a specific phase applies to all individual companies of an industry.

Sölvell, Zander and Porter (1991) further underscore a predictable ambition and strive for explanations for competitive behavior:

> No theory can fully explain the evolution of internationally competitive firms as a fully deterministic process. Chance events sometimes come into play. A close look at internationally successful and unsuccessful firms and industries reveals a complex process in which many influences play a role and shift in importance over time. However, the process by which competitive advantage is created is far from random. It arises from a predictable process in which national circumstances shape the perceptions of business opportunity and the tools and will to achieve it. (Sölvell, Zander, and Porter, 1991)

I agree that national circumstances shape the perceptions of business opportunity to some extent, but I argue that each industry and company has to be treated separately. Perceptions of opportunities differ from company to company, and even from one individual to another. Moreover, premises concerning technology are generally industry specific, and this means that the researcher has to be cautious before drawing conclusions valid for more than one industry. This need for cautiousness is also noted by Bohlin and Granstrand (1994) when they question the validity of traditional theory and models of internationalization for the internationalization of the telecommunications industry.

If we continue with the identification of previous empirical observations in the telecommunications industry, studies of deregulation have focused on major issues such as: (1) the roles of government or the regulator (Bergendahl-Gerholm and Hultkrantz, 1996, Burton, 1997, Prosser, 1997, Scott, 1996, Spiller and Cardilli, 1997, Trebing and Estabrooks, 1995, Waverman and Sirel, 1997), and (2) effects of deregulation on competition in specific countries, such as small innovative countries (Spiller and Cardilli, 1997), the U.S.A. (Shepherd, 1997), Japan (Fransman, 1997), European countries (Waverman and Sirel, 1997), the UK (Cave and Sharma, 1994), and Sweden (Granstrand and Johansson, 1994, Karlsson, 1998).

When it comes to roles of government and national regulators, the following major issues exemplify topics which have been studied:

- State of deregulation of countries in the European Union (Waverman and Sirel, 1997).
- Contrasting of the UK's model of utility, notably telecommunications services, deregulation with its working in practice (Burton, 1997).
- Selection of public policies that shape emerging infrastructures to meet societal efficiency, equity, and universal service goals (Trebing and Estabrooks, 1995).
- Effects of regulatory decisions such as specified interconnection rules, equal access to local networks, restriction of the number of network operators (Spiller and Cardilli, 1997).
- Consequences of price regulations (Bergendahl-Gerholm, 1996).

Regarding deregulation developments, Waverman and Sirel clearly showed the importance of national circumstances and these authors concluded:

> Full competition and liberalization in telecommunications are still mostly words in much of Europe, as indeed they are in the United States where there is little competition in local service and the regional Bell operating companies are still prevented from offering in-region local distance service. The European Commission is the agent for change and most of what has occurred in Europe, outside a few countries like the United Kingdom, is due to the Commission. Yet the Commission and its antitrust laws do have a political side. (Waverman and Sirel, 1997)

The effects of deregulation on competition among network operators in specific countries have been observed by other authors as well, without, however, going deeply into firm strategy. For example, Shepherd (1997) applied an economic perspective and tried to determine criteria for effective competition based on an examination of competition in telecommunications in the U.S.A. Further, Fransman (1997) focused on Japan and raised the question as to how to endure the establishment of competitive conditions in markets for telecommunications services.

Spiller and Cardilli (1997) studied four countries with varied national circumstances but innovative regulatory reform efforts (Australia, Chile, Guatemala and New Zealand). Deregulation has had striking results on the price and quantity of service in all four nations, and the intensity of competition in local telephony services has been especially intriguing, particularly as it was long considered a natural monopoly.

Cave and Sharma (1994) observed the results of the UK governmental policy of 1991 to permit virtually any operator to offer service, using a wide range of local access technologies, including cable TV and wireless technologies. Cave and Sharma particularly highlighted the competitive struggle between British Telecom, Mercury (later a part of Cable & Wireless), and cable operators in both the residential and business markets.

As regards the Swedish telecommunications market, Karlsson (1998) described the liberalization process and, in particular, analysed the role of technology in the changing competition.

Further, Granstrand and Johansson (1994) observed the attraction of large international players due to the alleged precursory nature of the liberalized market. This together with its smallness and advanced nature, makes Sweden attractive for trial and error.

Other studies have focused on general competitive developments in the industry, with no specific treatment of deregulations. Porter (1990) included descriptions of the telecommunications industry in his study of several different industries, but the question here was to scrutinize national competitiveness based on industry studies. Johansson (1994) specifically scrutinized telecommunications operators in his survey study. Further, the studies of Valetti and Cave (1998) and Mölleryd (1997) exemplify developments in parts of the industry, notably mobile telephony in the UK and Sweden, respectively.

If we consider studies of firm strategy, it is examples of previously established telecom operators which have been highlighted: British Telecom (Beesley and Laidlaw, 1989, Dang-N'guyen and Phan, 1994, Williams and Taylor, 1994), Cable & Wireless (Kramer and NiShuilleabhain, 1994), Mercury (Beesley and Laidlaw, 1989), Nippon Telephone and Telegraph (Fransman, 1997), Telia of Sweden (Carleheden, 1999, Parasiris, 1995), US West and Bell South (Noda and Bower, 1996). These studies, however, focus mainly on single firms and lack broad industry descriptions.

Thus, we are now able to conclude that there is a lack of studies on strategy developments in the telecommunications operator industry, simultaneously covering competition given specific national circumstances, and behaviour of individual firms.

Generally, we need to know more about strategy development in markets that emerge as a result of such changing conditions as deregulations. How do companies establish themselves in markets in specific countries? In particular, how do they perceive relevant entrance barriers, such as regulations? What strategy competence do they possess? What entry strategy do they choose? Further, which initial business strategies do new entrants and previously established competitors follow? Moreover, how are organizational structures adjusted to fit strategy development? What may cause restructuring and accompanying integration efforts?

I argue that comprehensive descriptions of strategy development in, for example, the telecommunications industry, are necessary in order to answer such questions and to extend our general understanding of strategy in emerging markets. However, the possibilities to make general conclusions beyond the industry studied are limited, not least due to changing premises concerning technology.

Subject studied and structure of this book

Through descriptions of the industry of telecommunications operators in the UK and in Sweden, including multiple case studies, this book applies a market establishment model and the strategic states model of business strategies. Initially, these models are based on literature reviews, including consideration of general strategy concepts and some key correlates. The strategy concepts and the components of the models are used as a framework for empirical descriptions.

Descriptive findings are related to the models in discussion, and empirical patterns are detected. In addition, findings are viewed from a broader strategy perspective in order to delineate processes of corporate restructuring and integration that are key elements related to strategy development.

The operators in question are those who held licences as public telecommunications operators for stationary or mobile telephony in 1996, either in the UK or in Sweden. This means that operators offering telephony services without possessing network licences are not included.

Figure 1.2 shows the logic of the research presented in this book.

This first chapter discusses the need for knowledge, presents the actors and markets on which this book focuses, and summarizes the framework for the book. Further, this chapter gives a background to the empirical studies of the book in terms of the major causes for the emerging markets of telecommunications operators: deregulation processes and technology development.

Part I of the book (Chapters two and three) comprises theoretical considerations. Some key strategy perspectives and

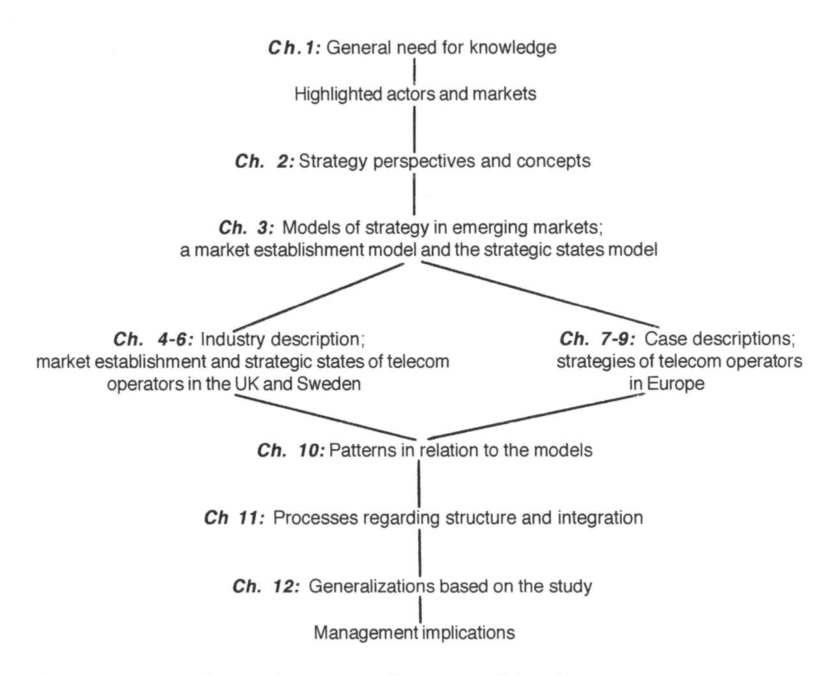

Figure 1.2 Linkages between chapters of this book.

concepts are discussed, and this results in the two models for strategy in emerging markets.

The empirical descriptions are presented in Parts II and III. A cross- sectional study of industry features in the UK and Sweden (Chapters four through six) follows the framework given by the models, and situations are illustrated with company examples. Furthermore, comprehensive multiple case studies give detailed information on strategy in emerging markets (Chapters seven through nine).

The final part of the book relates descriptive findings to the models and key strategy concepts, and accordingly discusses empirical patterns and processes (Chapters ten through eleven), making way for conclusions on strategy in emerging markets in general and on implications for management (Chapter twelve).

FRAMEWORK USED IN THE BOOK: A SUMMARY

I advocate that the strategy concept should include a definition of a product and the market scope of a company or individual businesses, and of the direction one wishes to follow given a long-term perspective. In practice, any company needs an explicitly planned strategy in order to survive under profitability constraints in competitive environments. Furthermore, my view is that a strategy frequently develops not in a straight line, but through a series of complicated variations. This means that retrospective analyses of strategy processes can be very valuable in formulating alternatives for the future.

I apply the perspective of strategic business units in which the formulation of corporate strategy concerns product and market issues relevant to an entire company group, while business strategy defines the choice of product, or service, and market of an individual business. Therefore, the following discussion treats corporate strategy, in terms of corporate offerings, and business strategy.

The framework for business strategy in emerging markets relies on concepts reflecting a company's process of establishment in an emerging market and the strategy the company caters for in the market in question.

Corporate offerings

Corporate strategy formulation and implementation are strongly correlated with issues pertaining to corporate structure and integration. For example, if there is an explicit desire to exploit common experiences and bring consistent offerings to the market, the forming of an appropriate organizational structure and implementing the accompanying integration measures are essential.

The corporation offers core competencies, core products, end products, core values, and a brand identity. I argue that these interrelated components of a corporate offering together constitute the "product" dimension of corporate strategy (Figure 1.3). In general, corporate integration facilitates the exploitation of core

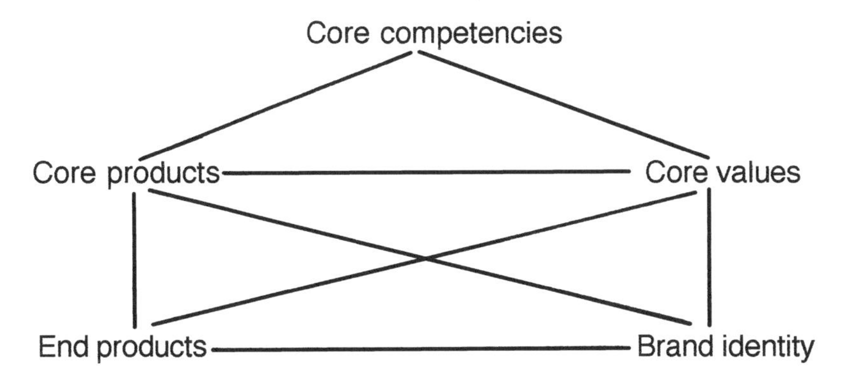

Figure 1.3 Components of a corporate offering.

competencies and their outgrowths. This is illustrated in the analysis of corporate strategies in the book.

A core competence provides potential access to a wide variety of markets. Further, a core competence should make a significant contribution to the perceived customer benefits of the end product. A core competence should also be difficult to imitate. A rival might acquire some of the technologies that comprise the core competence, but it will find it more difficult to duplicate the more or less comprehensive pattern of internal coordination and learning.

The tangible link between core competencies and end products is the core products. These are the physical embodiments of one or more competencies. Core products are the components or subassemblies that actually contribute to the value of the end products.

Besides core products, I argue that core values are linked to core competencies. Such values would reflect the values and history of the company and the values of the customers. Ideally, core values would define directions for product development, market communication, and customer interactions. This means that core values are reflected to varying degrees in physical core products and end products.

Those discussions that take place within a brand-oriented company concerning core values form the brand. Here, the intention would be to create a brand (consisting of functional, emotional and/ or symbolic attributes) that reflects the core competencies, products, and values of the corporation. Moreover, the brand should be perceived as unique by customers and by the organization itself, and the brand should be difficult to imitate by competitors.

Business strategy: the market establishment model

The interrelated establishment components form a model for market establishment (Figure 1.4). Here, the components concern the perceptions of relevant barriers to entering the market, the strategy competence sustaining the efforts, and the chosen strategy for entering the market.

In general, I regard barriers to entry as the major external factor influencing the outline of an entry strategy. The presence of barriers is the ultimate external factor that determines the likelihood of becoming firmly established in a market; however, such barriers are perceived differently from company to company and from one individual to another. This means that, regardless of the "objective" existence of barriers, different perceptions imply varying assessments of business opportunities. Such assessments decide what entry strategy is regarded as appropriate. Conversely, the process of implementing a specific entry strategy might result

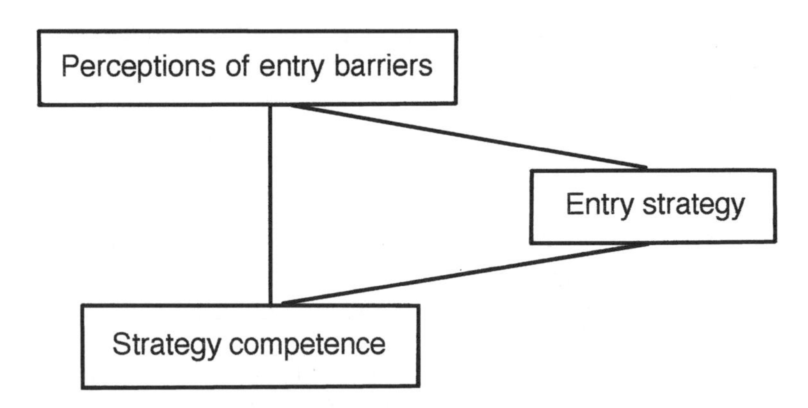

Figure 1.4 A market establishment model.

in changing perceptions as management learns about market conditions throughout the process; therefore, perceptions of entry barriers might change as a result of the implementation process.

In turn, both perceptions of entry barriers and knowledge gained during the process of implementing an entry strategy are strongly related to comprehending the need for internal strategy competence in order to become firmly established in the market. That is, management constantly needs to evaluate to what extent the required competencies are available.

In most cases, any company that wants to establish itself in a market has to face entrance barriers such as the need to acquire enough volumes to reach low costs, loyalties among buyers and previously established sellers, need for capital investments, costs for customers that want to switch from one supplier to another, limited access to distribution channels, and governmental policies.

The present study of network operators focuses on governmental regulations and the need for capital. Regulation aspects are captured by information on relevant perceptions, including views on the roles of the regulators, Oftel in the UK and The National Post and Telecom Agency in Sweden. Capital needs are discussed in terms of investments in the construction of infrastructures.

Broadly speaking, a market entry strategy requires decisions on factors that are relevant to entry into emerging markets:

- The objectives and goals in the target market.
- Needed policies and resource allocations.
- The choice of entry modes to penetrate the market.
- The control system to monitor performance in the market.
- A time schedule.

In this study, entry modes represent the broader term of entry strategies. A market entry mode is generally an institutional arrangement necessary for the entry of a company's products, technology, and human and financial capital into a market. Information on organic developments of sole ventures and of acquisitions, and on strategic alliances reflects modes of entering the emerging markets.

As I view it, the likelihood of becoming firmly established in a market depends, to a large extent, on the relevant strategy competence of the company. That is, the higher the competence, the better the general likelihood of becoming firmly established in the market.

I argue that the strategy competence of a certain company is due to the position of its businesses in relation to the core business of the corporation ("relatedness"). This means that efforts to establish a business that may be considered a part of the original corporate core business hypothetically have the best chance of being successful. The opposite is valid for a business unit that is less related to the core of the corporation.

Besides relatedness, the term "strategy competence" includes experience in the market. Familiarity with market conditions clearly strengthens competence and the likelihood of finding a sustainable position in the market. Thus, increasing knowledge of products, services, and other market conditions give extended market experience.

In the present study, information on the pertinent relationships between telecom operations and the original corporate core businesses indicate the relatedness aspect of competence. Market experience is reflected by the point in time at which the market in question was entered. Operators entering the UK and Swedish markets before 1991 are referred to as "early entrants", while those entering after that year are called "late entrants". The reason for the choice of 1991 as the breaking point is that the duopoly of operators, consisting of British Telecom (BT) and Mercury in the UK in that year was deregulated, and regional licences were granted to other operators and cable companies. The number of competing operators in Sweden also increased in 1991.

Business strategy: the strategic states model

The initial strategy in the market is captured by the strategic states model (Figure 1.5), which I have developed and validated in a number of studies. The environment of a company is reflected in the dimension of the number of market segments that are being

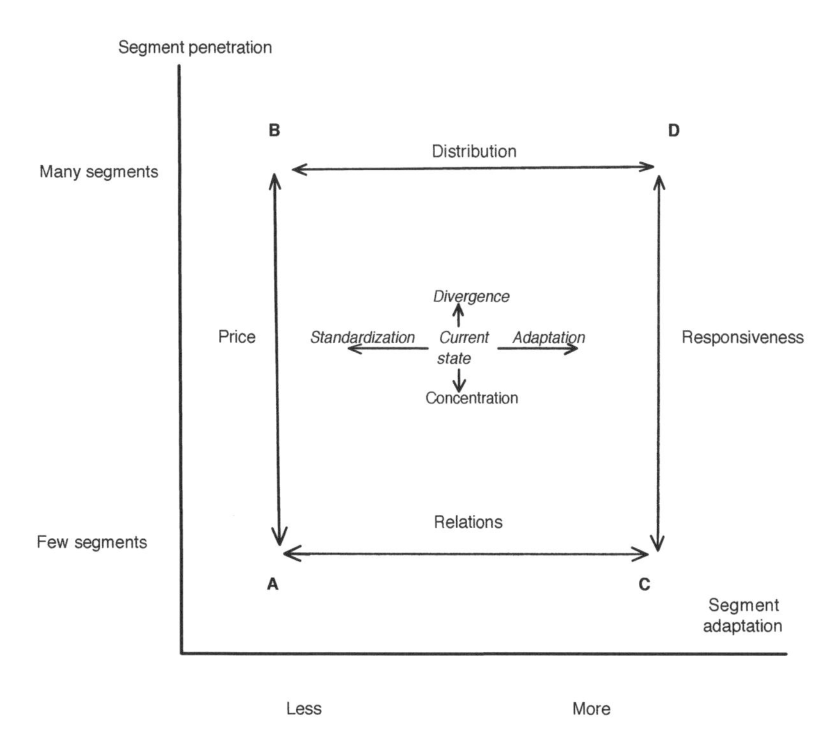

Figure 1.5 The strategic states model.

focused on, in which a segment is a limited and measurable part of a larger market. The other dimension is valid for the organization itself and concerns the adaptation of the offer to the requirements of the customers in the market segments. Thus, this model enables us to describe the strategic development of any business unit and its current strategic state, as well as alternatives for the future. Divergence, concentration, standardization and adaptation represent pure strategic alternatives.

Moreover, the degree of the unit's segment penetration and adaptation decides its sensitivity to variations in price and other competitive means, as well as the possibilities of treating changes in competition or other environmental fluctuations. Hence, the freedom of action varies according to the two dimensions of the model. In each state there is an optimum business strategy to reach high performance, and this strategy emphasises a specific combination of competitive edges.

The states of A–D in Figure 1.5 demonstrate the need for different combinations of competitive edges:

- Standardization of products and penetration of a few market segments (state A) imply that low prices and relationships building are central features of the optimum strategy.
- Standardization of products and penetration of many market segments (state B) imply that low prices and efficient distribution are central features of the optimum strategy.
- Adaptation of products and penetration of few market segments (state C) imply that responsiveness and relationships building are central features of the optimum strategy.
- Adaptation of products and penetration of many market segments (state D) imply that responsiveness and efficient distribution are central features of the optimum strategy.

In the application of the strategic states model in this book, classifications of competing operators are made according to the state appropriate for each operator. Thus, clusters of business strategies that competing companies follow are identified. Further, operator strategies within the framework of the model are delineated, and competitive edges are identified. Companies and private consumers represent customers that operators cater for, while the degree of segment adaptation is manifested by the type of end-product (telephony, or broader communications services including telephony), representing various degrees of customization.

EMERGING MARKETS FOR TELECOM OPERATORS IN EUROPE

A number of processes are taking place which are gradually changing the premises for competition for network operators. Far-reaching deregulation processes are in progress that are opening up the markets for new entrants and are extending the supply. Development of transmission technology and mobile telephony are other major changes that have significant impact on the markets. These changes will be outlined in order to give a background for the empirical studies of this book.

Deregulation processes

Telecommunications constitutes an important element in each country's infrastructure and is generally provided under governmental auspices or through government- controlled administrations. This is also why competitive rules for telecommunications are decided, to a large extent, by governments and parliaments. In all countries, there are restrictions on the types of services that are permitted to use the national telecommunications networks.

However, the earliest telephony operators were normally controlled by private interests. From 1880 to 1920, development of the industry was primarily based on competition between independent private operators.

As there was a need for uniform national standards and as it was deemed politically suitable, state-owned organizations were created in order to control national networks. Thus, such monopoly organizations controlled telephony operations during the subsequent decades in most countries. Shifting political attitudes and technological advances at the end of the 1970s then set the foundation for the forthcoming processes of deregulation of activities of operators (see e.g. Karlsson, 1998, for a review of deregulation processes).

In Europe, Japan, and North America, numerous discussions on degulation have taken place. When it comes to the European Union, the so-called Green Paper of 1987 set the political foundation for further deregulation. This formal agreement covered principles such as the following:

- The installation of regulating authorities, independent of governments.
- The availability of stationary networks controlled by the earlier monopoly organizations.
- Liberalization of telecommunications services, except traditional voice telephony.

- Liberalization of the terminal market in the widest sense, including all customer equipment.

In July 1996, deregulation was extended to infrastructure technology based on alternatives to the existing stationary networks. This meant that capacity in substitute networks such as power networks, cable TV networks, and rail networks were available for telecommunications purposes.

Free competition for all services and infrastructures became a goal, and the intention, in principle, to abolish the legal monopolies of national public telecommunications administrations in 1998 was explicitly expressed. Thus, through several legislative acts, the European Union had, to a large extent, accomplished these policies on the formal level by the middle of the 1990s.

The implementation of the new policies involved sometimes intense debates on practical issues. In the middle of the 1990s, three topics dominated the discussion within the Union: assurance of universal service obligations (that is, services available for all users), determination of terms for licences, and aspects related to connections between single operator's networks, including interconnection fees.

The UK experience of privatization and liberalization has been important for policy decisions. The UK was the first European country to begin a comprehensive process of deregulation. In 1981, telecommunications activities were separated from the other activities of the post and telecommunications administration, making way for the establishment of British Telecom (BT). Later, market liberalization made it possible for several companies of different origins to enter the operator market in the UK.

Unlike many other European countries, there has never been a legal monopoly in Sweden for the establishment of telecommunications networks or for the offering of services. However, Televerket (the Swedish public telecommunications administration) historically had a monopoly-like hold on many sectors of the industry. This organization was converted in 1993 into a commercial group with a parent company, Telia, and subsidiaries.

As there are no regulations protecting Swedish interests or restricting foreign operators from establishing themselves in the country, many companies have entered the market.

Technology development

Telecommunications equipment and services, maintenance, support, and office equipment are normally referred to as the foundation of the telecommunications industry. Telecommunications equipment consists of telephones, switching equipment, transmission equipment, receiver terminals, and other equipment. The number of telecommunications services (call charges, leased lines, switching services, and so on) is constantly increasing. Simultaneously, traditional distinctions between services are disappearing as integrated telecommunications usage transcends previous boundaries (see e.g. Mansell, 1993, for a discussion on technology convergence). In fact, technology convergence gives rise to questions of limitations in a study like this.

In this study, telephony covers local calls, domestic long-distance calls, international calls, and calls within single companies or organizations. Telephony and broader communications services may use either stationary or mobile networks, although the division between stationary networks using cables and mobile networks using radio technology has become less clear. Further, these services may also be integrated with Internet services using stationary and/or mobile networks.

Besides subscriber telephones, a mobile telephony network consists of radio-based stations that communicate with the telephones using antennas located in elevated situations. These stations are normally linked to switching systems by different types of cables.

In principle, three technologies exist for the transmission of communications signals: traditional copper cables, fiber optic cables, and radio transmission. The national networks of the previous monopolies, which have been built up over decades, primarily consist of copper cables. The fiber optic technology uses a much higher cable capacity at a lower cost, which means that network costs tend to be less dependent on geographical distances. This is particularly obvious in international traffic, where competition among network operators has become intense.

If an operator wants to offer international telecommunications to customers with extensive communications needs, it is, in principle, enough to supply a local loop that connects each customer to the long-distance and international lines directly. The operator might own these lines itself or rent them from another network operator.

High costs for cables and other equipment mean that households and small companies are normally not profitable when it comes to direct connections; therefore, there is a need for a so-called local access network. The traditional technology for local access is based on copper cables. As operators such as BT in the UK and Telia in Sweden possess established local networks, new operators in national markets are, to a large extent, dependent on interconnection and use of these existing local networks.

Alternatives exists, however, for local access to households and small companies. For example, cable TV companies may exploit their networks in order to offer telecommunications services as complements to original services. Radio technology is another option in local access networks. This citation by the regulator reflects the situation in the UK:

> BT is only now beginning to face serious competition in the access network from some cable companies and other alternative access suppliers, while radio-based competition has yet to begin service. There is lively competition in the UK market as a whole. It is growing and thriving. But it is not yet sufficiently effective so that market forces and general competition law could start to take over from specific regulation. (Oftel Statement, 1995)

The development of the cellular radio capable of accessing the public telephony network was a turning point in the history of telephony. For the first time, subscribers were not tied to any particular location to make and receive calls. However, during the first years, different systems were constructed according to various international standards. This made communication between the existing systems practically impossible.

The only exceptions were the Nordic countries (Denmark, Finland, Norway, and Sweden), where the Nordic Mobile Telephony (NMT) system, put into operation in 1981, provided cordless mobile telephone communications in an analog manner between mobile subscribers in the four countries. In 1994, this system was operational in 28 countries (Teldok Info, May, 1994), and it was followed by competing standards in other countries.

In order to set up a uniform standard for mobile telephony in Europe, using the digital technique, a number of countries agreed in the 1980s to choose the Global System for Mobile Communications (GSM) standard. In fact, the issue of standardization is common in the development of not only mobile telephony but stationary telephony.

PART I

Theoretical Considerations

INTRODUCTION

This part of the book considers theoretical concepts and models crucial to the study of strategy in emerging markets. Chapter 2 comprises a review of the literature on strategy and some key correlates: the strategy process, structure, company integration, and performance. I also develop my own thoughts on these important issues.

Based on the introduced theory framework, Chapter 3 presents two models for the study of strategy in emerging markets. One model treats the establishment process, while the other concerns current strategy in emerging markets.

TWO

Strategy and Some Key Correlates

This chapter gives an account of discussions on the strategy concept and on key correlates that are essential within the context of strategy in emerging markets: the strategy process, structure, integration, and performance. To begin, we need to know what strategy means and how it develops. Second, as strategy is formulated for organizations, we need to consider key organizational features, such as structure and company integration. Finally, as every strategy has to be financially defensible in the long term, it is important to be aware of general relationships between strategy and performance.

Regarding strategy and the strategy process, I make the following arguments:

- The strategy concept should include a definition of a product and market scope of a company or individual businesses and of the direction one wishes to follow given a long-term perspective.
- A strategy frequently develops not in a straight line but through a series of complicated variations.
- Retrospective strategy analyses can be very valuable in formulating alternatives for the future in order to delineate the emerging limits for freedom of action.
- In practice, any company needs an explicitly planned strategy in order to survive under profitability constraints in competitive environments.
- For research purposes, descriptive data on strategy development needs to be classified in accordance with a framework.

When it comes to strategy and structure, I suggest that core competencies should generally be the responsibility of the corporate level in an organizational structure consisting of strategic business units. Further, the strategy and integration discussion leads to two conclusions: (1) core competencies, core products, end products, core values, and brand identity constitute the "product" dimension of strategy, and (2) in general, company integration facilitates the exploitation of core competencies and their outgrowths, as well as the marketing of consistent offerings.

Finally, I come to the conclusion that it is possible to establish relationships between strategy and performance; that is, strategy differences may lead to performance consequences.

STRATEGY AND THE STRATEGY PROCESS

While the need for strategy concepts has developed from management practice, much of the elaboration and refinement of these concepts has occurred in the management literature. Drucker (1954) was among the first to address the strategy issue, although he did so only implicitly. To him, an organization's strategy was the answer to the dual question: "What is our business? What should it be?"

After Drucker's initial statement, little attention was given to the concept of strategy in management literature until Chandler, a business historian, published his seminal work in which he defined strategy as

> the determination of the basic long-term goals and objectives of an enterprise, and the adoption of courses of action and the allocation of resources necessary for carrying out these goals. (Chandler, 1962)

It is clear from this definition that Chandler did not differentiate between the process used to formulate the strategy and the concept itself. This was not a major problem for him, however, since his main interest was in studying the relationships between the way firms grew (their strategies) and the pattern of organization (their structures) devised to manage such growth.

Hofer and Schendel (1978) state that the first two authors who focused explicitly and exclusively on the concept of strategy and on the process by which it should be developed were Andrews (1965) and Ansoff (1965). Andrews combined both Drucker's and Chandler's ideas in his definition of strategy. For him,

> strategy is the pattern of objectives, purposes or goals and major policies and plans for achieving these goals, stated in such a way as to define what business the company is in, or is to be in, and the kind of company it is, or is to be. (Andrews, 1965)

Ansoff, in contrast, viewed strategy as the "common thread" among a company's activities that defines the essential nature of its business, both in the present and in the future. Ansoff then went on to identify four components that such a common thread would possess. These are:

- A product and market scope (the products offered by the company and the markets the company is in, that is, the company's entire business).
- A growth vector (the changes the company plans to make in its product and market scope: concentration, differentiation, diversification, and/or integration of resources).
- Competitive advantages (those particular properties of individual businesses that give the company a strong competitive position).
- Synergy (measures of joint effects between individual businesses).

Andrew's and Ansoff's discussions of strategy and the strategy formulation process, Hofer and Schendel continue, differed in three major points:

- The breadth of the strategy concept.

  Here, the question was whether the concept included both the goals and the objectives the company wishes to achieve and the means that will be used to achieve them (Andrew's view), or whether it included only the means (Ansoff's view). In other words, the broad versus the narrow concept of strategy.

- The components of strategy, if any.
  Here, the question was whether the narrow concept of strategy has components (Ansoff said yes, Andrews no), and if so, what they were.
- The inclusiveness of the strategic planning process.
  Here, the question was whether goal setting is a part of the strategic planning process (Andrews said yes), or whether it is a separate process (Ansoff's view).

In the years since Andrews and Ansoff presented their initial concepts of strategy and models of the strategic planning process, numerous authors have written on the topic, including Andrews (1987), Barnett and Burgelman (1996), Chakravarthy and Lorange (1991), Hamel and Prahalad (1993), Hofer and Schendel (1978), Lorange (1984), Mintzberg (1990a), Mintzberg and Waters (1985) and Porter (1980 and 1985).

The prescriptive design of strategies suggested by authors such as Andrews (1965 and 1987), Ansoff (1965), Chakravarthy and Lorange (1991) Hofer and Schendel (1978), Lorange (1984) and Porter (1980 and 1985) is the dominant view of strategy. Here, strategy formulation is essentially treated as a process of conceptual design, formal planning, and analytical positioning.

Another major treatment of strategy and the strategy process is the descriptive school, in which Chandler (1962) pioneered by studying relationships between strategy and structure over long time periods. Later, Mintzberg (1990a) and Mintzberg and Waters (1985) took the view that strategy formation is an emergent process. Further, the evolutionary perspective is based on descriptions, and the aim here is to develop dynamic path-dependent models that allow for possible random variation and selection within and among organizations (Barnett and Burgelman, 1996).

Andrews' view of strategy clearly belongs to the prescriptive school. In 1987, he developed his position and made the following argument:

> Corporate strategy is the pattern of decisions in a company that determines and reveals its objectives, purposes, or goals, produces the principal policies and plans for achieving those

goals, and defines the range of business the company is to pursue, the kind of economic and human organization it is or intends to be, and the nature of the economic and non-economic contribution it intends to make to its shareholders, employees, customers, and communities. (Andrews, 1987)

Andrews went on to say that in an organization of any size or diversity, corporate strategy usually applies to the whole enterprise, while business strategy, which is less comprehensive, defines the choice of product or service and market of individual businesses within the firm. That is, business strategy is the determination of how a company will compete in a given business and position itself among its competitors. Corporate strategy defines the businesses in which a company will compete, preferably in a way that focuses resources to convert distinctive competence into competitive advantage. Both are outcomes of a continuous process of strategic management.

The strategic decision contributing to this pattern is one that is effective over long periods of time, affects the company in many different ways, and focuses and commits a significant portion of its resources to the expected outcomes. The pattern resulting from a series of such decisions will probably define the central character and image of a company, the individuality it has for its members and various publics, and the position it will occupy in its industry and markets. It will permit the specification of particular objectives to be attained through a timed sequence of investment and implementation decisions and will govern directly the deployment or redeployment of resources to make these decisions effective.

Some aspects of such a pattern of decisions may be seen in the ways an established corporation remains unchanged over long periods of time, such as in a commitment to quality, to high technology, to certain raw materials, or to good labor relations. Other aspects of a strategy must change as or before the world changes, such as a product line manufacturing process or merchandising and styling practices. The basic determinants of company character, if purposefully institutionalized, are likely to persist through time and to shape the nature of substantial changes in product-market choices and in allocation of resources.

> The essence of the definition of strategy I have just recorded is pattern. The interdependence of purposes, policies, and organized action is crucial to the particularity of an individual strategy and its opportunity to identify competitive advantage. It is the unity, coherence, and internal consistency of a company's strategic decisions that position the company in its environment and give the firm its identity, its power to mobilize its strengths, and its likelihood of success in the marketplace. It is the interrelationship of a set of goals and policies that crystallizes from the formless reality of a company's environment a set of problems an organization can seize upon and solve. (Andrews, 1987)

Porter (1980) also takes a prescriptive view of strategy and talks about two main directions for strategies: a search for integration to achieve cost-leadership and a search for differentiation. Each of these directions may be valid for a broad market or for a specific part of the relevant market. At any rate, a commercial organization must face a combination of these strategic features in any given situation (Lorange, 1984). The relative importance of each feature will be dictated by the environmental setting at hand and will tend to differ from business to business and from company to company. It is likely to change over time as well.

As Andrews is one of the leaders of the prescriptive design school, he receives some criticism:

> Andrews thought it sufficient to delineate one model and then add qualifications to it. The impression left was that this was the way to make strategy, although with nuance, sometimes more, sometimes less. But that had the effect of associating strategy making with deliberate, centralized behavior and of slighting the equally important needs for emergent behavior and organizational learning. Another extreme, the "grassroots model," makes no more sense, since it overstates equally. But by positioning these two at ends of a continuum, we can begin to consider real-world needs along it. (Mintzberg, 1990b)

As the citation above indicates, Mintzberg rejects prescriptions and instead suggests that strategy should be viewed as a pattern in a stream of decisions taken in the strategy formation process. Thus, Mintzberg represents the descriptive way of viewing strategy.

Mintzberg and Waters (1985) have been researching the process of strategy formation based on the definition of strategy as a pattern in a stream of decisions. They argue that streams of behavior could be isolated and strategies identified as patterns or consistencies in such streams. The origins of these strategies could then be investigated, with particular attention paid to exploring the relationship between leadership plans and intentions ("intended strategy") and what the organizations actually did ("realized strategy"). Comparing intended strategy with realized strategy allowed Mintzberg and Waters to distinguish deliberate strategies, realized as intended, from emergent strategies (that is, patterns or consistencies realized despite, or in the absence of, intentions).

The fundamental difference between deliberate and emergent strategy is that, whereas the former focuses on direction and control, the latter opens up the notion of strategic learning. Defining strategy as intended and conceiving it as deliberate, as has traditionally been done, effectively precludes the notion of strategic learning.

Emerging strategy itself implies learning what works, taking one action at a time in search for that viable pattern or consistency. It is often through the identification of emergent strategies that managers come to change intentions. Whereas more deliberate strategies tend to emphasize central direction and hierarchy, the more emergent ones open the way for collective action and convergent behavior.

The conclusion of Mintzberg and Waters is that strategy formation walks on two feet, one deliberate, the other emergent. Managing requires a light, deft touch to direct in order to realize intentions while at the same time responding to an unfolding pattern of action. The relative emphasis may shift from time to time but not the requirement to attend to both sides of the phenomenon.

Ansoff (1991) criticizes the strategy view of Mintzberg and Waters along the following lines. First, Ansoff mentions the self-denial of a chance to study the business environment:

> In turbulent environments, the speed at which changes develop is such that firms which use the emerging strategy formation advocated by Mintzberg endanger their own survival. When they arrive on a market with a new product or service, such firms find the market preempted by more foresightful competitors who plan their strategic moves in advance. (Ansoff, 1991)

Ansoff then distinguishes between different definitions of strategy:

Mintzberg's strategy definition is descriptive, since in order to identify the strategy it is necessary to wait until a series of strategic moves has been completed. But the concept used in practice is prescriptive and stipulates that strategy should be formulated in advance of the events that make it necessary. Mintzberg fails to differentiate between descriptive and prescriptive statements. (Ansoff, 1991)

The evolutionary perspective of the strategy process belongs to the descriptive view, although one aim is to present prescriptive models. Barnett and Burgelman (1996) claim that those strategy researchers who take an evolutionary perspective on strategy explicitly question how strategic outcomes develop. In summary, taking this perspective means developing dynamic path-dependent models that allow for possible random variation and selection within and among organizations. However, the problem of defining what constitutes evolutionary theory in general is not resolved. In the social sciences, most working definitions include the use of explicitly dynamic models and an allowance for randomness, variation, selection, and, sometimes, retention.

To contribute to the evolutionary perspective, Barnett and Burgelman continue, it is not necessary for a study to satisfy all the components of the definition. Most careful researchers look at only one or another aspect of strategic evolution, such as when a study looks only at failure rates or only at variations due to innovation, but all work in this vein studies strategic dynamics.

Here, specifying a dynamic model means constructing theory that can predict patterns of change, including rates of change and alternative paths of change (particular sequences of events). Dynamic models may predict convergence toward a steady state, several possible states, or possible ranges rather than states, but regardless of their treatment of equilibrium conditions, evolutionary models attend to the pace and path of strategic change. For instance, we might model how quickly and along which paths organizations will grow, change their performance, or experience strategic events such as birth, restructuring, product innovation, merger, technological change, or failure.

An evolutionary perspective allows for variation in the possible strategies that organizations pursue. But how do new strategic variants develop? How do organizations search for and learn about strategic options? These questions invite us to describe the rate and path of innovation among and within existing organizations, either when organizations grow or when strategic initiatives are launched within firms. In either case, an evolutionary perspective allows for many variations to arise essentially at random.

Furthermore, evolutionary inquiry asks how selection processes affect, and are affected by, the pace and path of strategic change. Selection processes often do not function as a smoothly and rapidly optimizing force. In order to understand strategic success, Barnett and Burgelman conclude that we must study both the winners and the losers, as we do in the systematic analysis of organizational failure.

Hamel and Prahalad (1993) try to create a bridge between the view of strategy as a plan (the prescriptive view), and strategy as a descriptive pattern. They claim that the answer to the question, What is strategy? from a practical point of view, usually centers on three elements:

- The concept of fit or the relationship between the company and its competitive environment.
- The allocation of resources among competing investment opportunities.
- A long-term perspective in which "patient money" figures prominently.

From this perspective, "being strategic" implies a willingness to take the long- term view, and "strategic investments" are those that require a large and preemptive commitment of resources as well as a distant return and substantial risk. Hamel and Prahalad think that this dominant strategy framework is not wrong, only unbalanced:

The predominance of these planks has obscured the merits of an alternative frame in which the concept of stretch supplements the idea of fit, leveraging resources is as important as allocating them, and the long term has as much to do with consistency of effort and purpose as it does with patient money and an appetite for risk. (Hamel and Prahalad, 1993)

Hamel and Prahalad suggest that the products of stretch are, for example, a view of competition as encirclement rather than confrontation, an accelerated product-development cycle, tightly knit cross-functional teams, a focus on a few core competencies, strategic alliances with suppliers, or programs for employee involvement and consensus.

Stretch can beget risk when an arbitrarily short-term horizon is set for long- term leadership goals. Impatience brings the risk of rushing into markets not fully understood, ramping up R&D spending faster than it can be managed, acquiring companies that cannot be digested easily, or rushing into alliances with partners whose motives and capabilities are poorly understood. The job of top management is not so much to stake out the future as it is to help accelerate the acquisition of market and industry knowledge.

The notion of strategy as stretch helps to bridge the gap between those who see strategy as a grand plan thought up by great minds and those who see strategy as no more than a pattern in a stream of incremental decisions. On the one hand, strategy as stretch is strategy by design, in that top management has a clear view of the goal line. On the other hand, strategy as stretch is strategy by incrementalism, in that top management must clear the path for leadership meter by meter. In short, strategy as stretch recognizes the essential paradox of competition: leadership cannot be planned for, but neither can it happen without a grand and well-considered aspiration. (Hamel and Prahalad, 1993)

Allocating resources across businesses and geographical regions is an important part of top management's strategic role, but leveraging what a company already has rather than simply allocating it is a more creative response to scarcity. In the continual search for less resource-intensive ways to achieve ambitious objectives, leveraging resources provides a very different approach from the downsizing and delayering, the restructuring and retrenchment, that have become common as managers contend with rivals around the world who have mastered the art of resource leverage.

There are two basic approaches to garnering greater resource productivity, whether those resources be capital or human. The first is downsizing, cutting investment and head count in the hope of becoming lean and mean. The second approach, resource leveraging, seeks to get the most out of the resources one has. Resource leveraging is essentially energizing, while downsizing is essentially demoralizing. Both approaches will yield gains in productivity, but a company that continually ratchets down its resource base without improving its capacity for resource leveraging will soon find that downsizing and restructuring become a way of life.

Hamel and Prahalad conclude by arguing that management can leverage its resources, financial and nonfinancial, in five basic ways: by concentrating them more effectively on key strategic goals; by accumulating them more efficiently; by complementing one kind of resource with another to create higher-order value; by conserving resources wherever possible; and by recovering them from the marketplace in the shortest possible time.

Based on the literature review above, I will now present my own thoughts on strategy and the strategy process. To begin, I adopt the narrow strategy concept of Ansoff (1965), where strategy is the means applied by the company to avoid environmental threats, to exploit environmental opportunities, given goals and internal resources and capabilities, and to survive in the long run under profitability constraints. The primary objective for survival is thus profitability.

From my point of view, the strategy concept includes a definition of a product and market scope of the company, or of the individual businesses, and a change direction one wishes to follow given a long-term perspective. Hence, I include neither competitive advantages nor synergy in the concept. Although it may be discussed whether the choice of direction or advantages comes first in practice, I regard the latter as dependent on the choice of long-term direction. Ambitions to achieve synergies belong to the important organizational questions crucial to implementing a chosen strategy.

This means that even if strategy comprehends "stretch" and the leveraging of resources (Hamel and Prahalad, 1993), a change direction has to be chosen. And this is my main argument against this view: stretch and leverage primarily have to do with how a certain strategy will be implemented. This view focuses on those aspects of resources that are essential in strategy implementation.

Further, goals are, in my opinion, formulated in the strategic planning process. They are the ultimate, long-run, open-ended attributes or ends the company seeks, while its objectives are the intermediate-term targets that are necessary but not sufficient for the satisfaction of goals (Hofer and Schendel, 1978). The structure of the objective ideally consists of an attribute (such as growth), an index (unit sales), a target (x units), and a time frame (three years).

If we extend the process perspective on strategy development, it is clear that a strategy frequently develops not in a straight line but through a series of complicated variations that often appear anything but obvious to the chief actors. Considering that it can take many years from the initial idea to a significant volume for a specific business and that judgments are going on continuously, a strategy is often constantly revised and developed.

A comparison with Kissinger's discussion on national strategies may shed light on the nature of strategic decision making:

> In retrospect, a single choice may seem to have been nearly random, or else to have represented the only available alternative. In either case, the choice is not an isolated act but an accumulation of previous decisions reflecting a history of tradition and values as well as the immediate pressures of the need for survival. (Kissinger, 1977)

Mintzberg and Waters (1985) derive a strategy by identifying the pattern of a company's previous decisions and resource allocations. Strategy is hence being inferred *ex post facto* by analyzing and understanding how resources have been allocated. Yet, an obvious question is to what extent understanding of how strategy has developed helps management make better decisions in the present and take more appropriate actions, following the intended strategy. The usefulness for practitioners to regard strategy as a pattern in a stream of decisions primarily lies in gaining a better understanding of how resources have been committed and, hence, being better able to manipulate the decision process.

Moreover, as I view it, if a better understanding of history and of the emerging limits for the freedom of action is available, this might help us to better appreciate the present and more correctly foresee the future. On the other hand, history is often used to rationalize decisions taken, or to justify mistakes. Since historical patterns and specific circumstances for decisions can frequently be generalized to any period of time, however, retrospective analyses can be very valuable when formulating alternatives for the future.

For example, a company competing in an industry where it is necessary to make a significant investment at a specific point in time has to be conscious of the resulting restriction on the freedom of action. A manufacturing company that invests in a new paper machine producing a certain paper quality (as an example) would find that the number of available strategy alternatives for the future would diminish dramatically because of the high fixed costs. For this company careful formulation of an intended strategy would be crucial, as an unrealized strategy will cause serious consequences. That is, it is a prerequisite that the intended strategy will turn into a realized strategy.

Another manufacturing company that has continually standardized its products and has become a manufacturer of electronics components, despite the emerging strategy perhaps would still have the opportunity to choose from several strategy alternatives. At the same time the industry as a whole would demonstrate overcapacity, resulting in stiff price competition. The company in question then would intend to create a well-defined profile and an explicit strategy for the future.

These examples show that Mintzberg and Waters (1985) fail to distinguish between different company situations and their impact on strategy. That is, the importance of intended and emergent strategies generally differs from one company situation to another. In practice, every company needs an explicitly planned strategy in order to survive in competitive environments, no matter how the previous strategy has developed.

Furthermore, the examples indicate that reliance on chance as the major determinant of strategy development, as advocated by the evolutionary perspective on strategy, will be practically disastrous. Companies relying on random decisions clearly will find themselves in danger. In fact, planning capability distinguishes the human species from others, although chance is the essential determinant of evolution in biological systems other than human systems. Therefore, random decisions very seldom exist in companies. On the contrary, each choice is an outgrowth of previous decisions reflecting tradition and values as well as immediate pressures, and, moreover, there is an obvious need for prescriptions in practical strategy formulation.

Strategy research is not, however, the same as strategy formulation in practice. Research concerning strategy development requires a descriptive definition of strategy. If we use the descriptive definition of Mintzberg and Waters that "strategy is a pattern in a stream of decisions", we need to define what type of decisions we will focus on and what qualifies a stream as a pattern. Afterwards we have to assess each pattern and denominate it as a certain strategy.

I think that possible strategies, or long-term change directions, that are valid for a specific product and market scope have to be defined in advance. This means that descriptive data for research purposes should be related to these strategy definitions in order to ascertain which strategy a company or a single business follows. Empirical data thus need to be classified in accordance with a strategy framework.

STRATEGY AND STRUCTURE

Simon (1962) argues that theoretically we could expect complex systems, such as company organizations, to be hierarchies in a world in which complexity had to evolve from simplicity. The central theme that runs through Simon's remarks is that complexity frequently takes the form of hierarchy and that hierarchical systems have some common properties that are independent of the systems' specific content. Hierarchy is one of the central structural schemes that the architect of complexity uses.

By a hierarchical system, or hierarchy, Simon means a system that is composed of interrelated subsystems, each of the latter being, in turn, hierarchical in structure, until we reach some lowest level of elementary subsystem. In most systems in nature, it is somewhat arbitrary where we stop the partitioning and what subsystems we regard as elementary.

In the Chandlerian interpretation of organizational structures in companies, some basic orientations of operational structures were outlined: functions or departments, divisions (valid for products, customers or geographical regions), and combinations (matrix orientation). Moreover, a company can act as a parent company in a group structure. Following the hierarchical idea, subsidiaries occupy the lower-level positions in such a structure.

For larger groups, which contain several individual businesses, the Strategic Business Unit (SBU) concept has become widely utilized. A major reason for supplementing an operational structure is the need to coordinate businesses. In accordance with Hall (1978), General Electric was one of the pioneers in implementing the SBU concept:

> It started in 1971, in the executive offices at General Electric (GE), the world's most diversified company. Corporate management at GE had been plagued during the 1960s with massive sales growth, but little profit growth. Using 1962 as an index of 100 dollars, sales grew to 180 by 1970. However, earnings per share fluctuated without growth between 80 and 140, while return on assets fell from 100 to 60. Thus, in 1971, GE executives were determined to supplement GE's vaunted system of management decentralization with a new, comprehensive system for corporate planning. The resulting system was based upon the new structure concept of strategic business units. (Hall, 1978)

In the SBU framework, corporate strategy usually applies to an entire company group, while business strategy defines the choice of product or service and market of an individual business (Andrews, 1987). Strategically distinct business units are isolated (Porter, 1985) by weighing the benefits of integration and de-integration and by comparing the strenghts of interrelationships in serving related segments, regions, or industries to the differences best suited for serving them separately.

Lorange (1980) and Chakravarthy and Lorange (1991) have refined the SBU concept and talk about hierarchies of strategies, which are to be elaborated within the hierarchical strategic structure. This structure consists of three major SBU levels: the corporate level, the business family level and the business element level. The number of SBUs on lower levels is not likely to remain constant over time. Furthermore, the strategic structure frequently differs from the operational structure, which in this context is built upon profit centers and has budgetary responsibility for executing strategies. Systematic observations have been made that the larger the organization, the more likely it seems to be to have SBUs that do not coincide with operating units (Haspeslagh, 1982).

On the corporate level, the main task, in theory, is to determine relationships among business families (Chakravarthy and Lorange, 1991). This can be examined in several ways. Questions that might be asked include: Is there a satisfactory mix of mature and new businesses and between high-risk/high-return, and low-risk/low-return businesses? What is the nature of the overall pattern of funds flow? How do the businesses support each other in terms of human resources, know-how, and distinctive expertise? An organization at this level is figuring out how the pieces all fit together.

Within each family on the business family level, strategies identify a meaningful set of product and market scopes, or niches, occupied by elements and determine possible synergies among them. For instance, what are the advantages in the marketplace of having a full range of products? Internally, one might examine whether there are facilities and resources that could be used by several business elements within the business family, such as common utilization of R&D, joint manufacturing, common sales organization, or coordinated logistics functions.

The primary task on the business element level is to develop clear competitive strategies for a particular product and market niche so as to beat specific competitors in reaching the targeted set of customers.

Strategic directions are quite often "fed into" the operational structure by so- called strategy teams who manage the SBUs. This indicates the strategic implementation tasks that each operating entity is expected to undertake in addition to its ordinary operating tasks.

There are, however, arguments against the usefulness of the SBU concept. For example, Prahalad and Hamel (1990) believe in a view of the company as a portfolio of competencies, besides portfolios of products and businesses:

> When you think about this reconceptualization of the corporation, the primacy of the SBU is now clearly an anachronism. Where the SBU is an article of faith, resistance to the seductions of decentralization can seem heretical. In many companies, the SBU prism means that only one plane of the global competitive battle, the battle to put competitive products on the shelf today, is visible to top management. (Prahalad and Hamel, 1990)

Prahalad and Hamel point out that no single business in a SBU structure may feel responsible for maintaining a viable position in core products or may be able to justify the investment required to build leadership in some core competencies. Further, as an SBU evolves, it often develops unique competencies. The people who embody this competence are typically seen as the sole property of the business in which they grew up. If core competencies are not recognized, individual SBUs will pursue only those innovation opportunities that are close at hand. Conceiving of the corporation in terms of core competencies widens the domain of innovation.

These arguments underscore, among other things, the importance of careful objective setting in relation to appropriate responsibilities of SBUs. In my opinion, profitability should be the major objective on as many strategy levels as possible, although practical efforts to break down this objective are frequently obstructed by difficulties in dividing responsibility for capital. Therefore, it may be more appropriate to control lower SBU levels by the use of profit margins or related objectives.

And here is the crucial problem: to what extent does an SBU, which has profitability responsibility, care about long-term aspects inherent in, for example, research on technology? Although I think that each company has to evaluate its own situation in this respect, core competencies should be assigned long-term perspectives and generally should be the responsibility of the corporate level.

STRATEGY AND INTEGRATION

There is no question that a company's competitiveness in the short run derives from the price and performance attributes of current products. In the long run, Prahalad and Hamel (1990) argue, competitiveness derives from an ability to build, at lower cost and more speedily than competitors, the core competencies that spawn unanticipated products. The real sources of advantage are to be found in management's ability to consolidate corporatewide technologies and production skills into competencies that empower individual businesses to adapt quickly to changing opportunities. This argument leads us to a discussion on integration which is a key correlate of strategy.

Prahalad and Hamel point out some features of core competence:

- It is the collective learning in the organization, especially how to coordinate diverse production skills and to integrate multiple streams of technologies.
- It is about the organization of work and the delivery of value. The force of core competence should be felt as decisively in services as in manufacturing.
- It is communication, involvement, and a deep commitment to working across organizational boundaries. It involves many people and all functions.
- It may guide patterns of diversification and market entry.

At least three tests can be applied to identify core competencies in a company. First, a core competence provides potential access to a wide variety of markets. Second, a core competence should make a significant contribution to the perceived customer benefits of the end product. Finally, a core competence should be difficult to imitate. Imitation will be difficult if it is a complex harmonization of individual technologies and production skills. A rival might acquire some of the technologies that comprise the core competence, but it will find it more difficult to duplicate the more or less comprehensive pattern of internal coordination and learning.

The tangible link between identified core competencies and end products is what Prahalad and Hamel call the core products (that is, the physical embodiments of one or more competencies).

Core products are the components or subassemblies that actually contribute to the value of the end products:

> It's essential to make this distinction between core competencies, core products and end products because global competition is played out by different rules and for different stakes at each level. To sustain leadership in their chosen core competence areas, the leading companies seek to maximize their world manufacturing share in core products. The manufacture of core products for a wide variety of customers yields the revenue and market feedback that, at least partly, determines the pace at which core competencies can be enhanced and extended. (Prahalad and Hamel, 1990)

I would like to add that, besides core products, core values are linked to core competencies of a company. Such values would reflect the values and history of the company and the values of the customers (Urde, 1997). Ideally, core values would define directions for product development, market communication, and customer interactions. This means that core values are reflected to varying degrees in physical core products and end products.

Those discussions that take place within a brand-oriented company concerning core values form the brand and start the integration of the brand and the strategy. The intention would be to create a brand that reflects the strategy, core competencies, and core products and that is perceived as valuable and unique by customers and by the organization itself. And the brand should be difficult to imitate by competitors.

A brand represents functional, emotional, or symbolic values. In general, emotional or symbolic values would provide the strongest protection of the brand identity. Further, the more the brand and the strategy as a whole are integrated, the more difficult it will generally be to imitate the brand identity.

So far we have discussed what the company offers: core competencies, core products, end products, core values, and a brand identity. I argue that these interrelated components together constitute the "product" dimension of strategy (Figure 2.1). In general, company integration facilitates the exploitation of core competencies and their outgrowths. A desire to present consistent offerings in a market favors integration, but, as the following discussion will show, there are also other causes of integration.

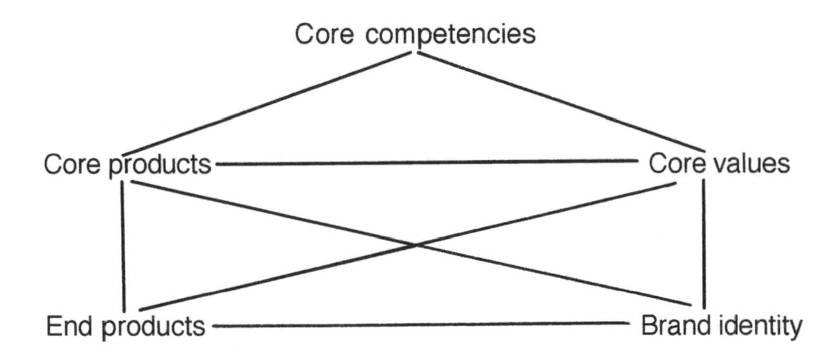

Figure 2.1 Components of a corporate offering.

A central question for a firm entering a new market is whether the new activities should be integrated into the firm's previous systems or whether the new activities should be left to operate as an independent autonomous unit. In an international context, this question was introduced by Perlmutter (1969) and Fayerweather (1969). Fayerweather provided a conceptual framework for analyzing the tension that prevails between the strategic choice of whether the management and strategy of subsidiary operations should be diversified according to individual requirements (fragmentation), or whether it should be an integrated part of the entire system (unification). "Unification" in this sense is synonymous with "integration" as it is commonly used in the literature, although Fayerweather states that integration does not necessarily encompass the standardization and uniformity that are intended in the unification idea.

The firm may favor integration for a number of reasons, which include the need for greater utilization of the firm's existing resources and capabilities through economies of scale, scope and learning. By integrating operations, the firm is able to utilize core competencies or other capabilities over a broader base. Unused resources and capabilities within the firm may be better utilized by expanding them into the subsidiary through integration, rather than by fragmentation.

Apart from efficiency rents, the firm is able to enhance learning among subsidiary units as a result of information exchange among network units through subsidiary integration. Integrating subsidiary operations allows a greater flow of information between subsidiaries in different market environments. Experiences arising from different local environments of subsidiaries can better be shared through information exchange, which is enhanced by greater integration of the subsidiaries. Thus, integration is likely to be the desired means for exploiting the firm's existing resources, as well as for developing the firm's core competencies through learning as a result of internal coordination among firm units.

Prahalad and Doz (1987) present a model that is important in the context of strategy and integration. Three building blocks are present in the model: global integration of activities, global strategic coordination, and local responsiveness. These blocks refer to the nature of relationships between headquarters and subsidiaries, as well as among subsidiaries in a multinational setting. However, those relationships are dependent on the nature of the business units in the diversified multinationals. The economic, technological and competitive characteristics of a business enable us to define pressures for global integration of activities and local responsiveness. The need for strategic coordination is harder to focus on. Typically, businesses that need significant global integration of activities also require strategic coordination; however, given active global competition, locally responsive business units may demand strategic coordination as well.

Integration in the context of Prahalad and Doz refers to the centralized management of geographically dispersed activities on an ongoing basis. Managing shipments of parts and subassemblies across a network of manufacturing facilities in various countries is an example of integration activity.

The need for integration in this model arises in response to pressures to reduce costs and optimize investments. Such pressures may force location of plants in countries with low labor costs. Products are then shipped from those plants to the established markets. The same pressures may also lead to building large, highly specialized plants in order to realize economies of scale. Managerially, this translates into a need for ongoing management of logistics that cuts across multiple national boundaries. These are some important pressures for global integration:

- Technology intensity and the extent of proprietary technology often encourage firms to manufacture in only a few selected locations. Having fewer manufacturing sites allows easier control over quality, cost and new product introduction. Centralized product development and manufacturing operations in a few locations results in global integration, particularly when the markets are widely dispersed.
- Pressure for cost reduction often implies global integration. Cost reduction requires sourcing the product from low factor cost locations (global sourcing), or exploiting economies of scale and experience by building large plants that serve multiple national markets. Either approach to lowering costs imposes a need for global integration.
- If the product meets universal needs and requires little adaptation across national markets, global integration is obviously facilitated.
- Access to raw materials and a cheap and plentiful supply of energy can force manufacturing to be located in a specific area.

Strategic coordination refers to the central management of resource commitments across national boundaries in the pursuit of a strategy. It is distinct from the integration of ongoing activities across national borders. Typical examples would involve coordinating research and development priorities across several laboratories, coordinating pricing to global customers, and facilitating transfers of technology from headquarters to subsidiaries and across subsidiaries. Unlike activity integration, strategic coordination can be selective and nonroutine.

The goal of strategic coordination is to recognize, build, and defend long- term competitive advantages. For example, headquarters may assign highly differentiated goals to various subsidiaries in the same business unit in order to develop a coherent response to competition. Strategic coordination, like the integration of activities, often involves headquarters and one or several subsidiaries. Coordination decisions transcend a single subsidiary.

Prahalad and Doz suggest that local responsiveness refers to resource commitment decisions taken autonomously by a subsidiary in response to primarily local competitive or customer demands. In a wide variety of businesses, there may be no competitive advantage to be gained by coordinating actions across subsidiaries. Typically, businesses for which there are no meaningful economies of scale or proprietary technology fall into this category. The need for significant local adaptation of products or differences in distribution across national markets may also indicate a need for local responsiveness.

STRATEGY AND PERFORMANCE

Since every strategy has to be financially defensible in the long term, it is important to be aware of general relationships between strategies and company performance. In order to facilitate the formulation of strategies in a given market, it may be helpful to examine the strategies of profitable companies.

Various universal models can be found in the strategic management literature prescribing, for example, high profitability for cost leaders and companies that follow the focus strategy (Porter, 1980). Other models rely on research that searches for relationships contained in broad categories, such as industrial and consumer goods companies (for example, Galbraith and Schendel, 1983, Rumelt, 1974 and 1982). Samiee and Roth (1992) espouse this tradition, comparing performance levels of companies offering standardized products with those offering less standardized products in a wide range of industries.

Most strategy researchers, in fact, offer some rationale to account for performance differences among organizations, or to account for strategic differences that, presumably, have performance consequences. Barnett and Burgelman (1996) give more examples of rationales for explaining the good performance of a company:

- It may be in a market position that is protected from competition.
- It may have unique capabilities that enable it to innovate or differentiate.
- It may occupy a powerful position in a network of organizations.
- It may have a strategy or structure that fits well with the challenges offered by the market.
- It may be efficiently designed so as to minimize transaction costs.
- It may have outwitted its rivals in strategic interaction.

A common belief among the examples mentioned is that a theoretical rationale can be expected to correspond to empirical patterns observable at any given time. In this belief, strategy researchers typically look for cross-sectional correlations in data at a single point in time, or sometimes even in a single case at a single time.

For example, Rumelt (1974) used cross-sectional data valid for several industries and found that corporate profitability differed significantly across groups of firms following different diversification strategies. The highest levels of profitability were exhibited by those having a strategy of diversifying primarily into those areas that drew on some common core skill or resource. The lowest levels were those of vertically integrated businesses and firms following strategies of diversification into unrelated businesses. In a complementary study (Rumelt, 1982), he observed a pattern quite similar to that observed in the original study. Vertically integrated and unrelated businesses showed levels of profitability significantly below those in the other strategy categories. As before, the most profitable group consisted of companies diversifying into related product areas.

Thus, the degree of diversification around a common core varies from firm to firm. Taking a business unit to be a product,

product line, or set of product lines that have strong market interdependencies, Rumelt discusses a firm's

- Specialization ratio (the fraction of revenues accounted for by the largest single business unit).
- Related-core ratio (the fraction of revenues attributable to the largest group of businesses that share or draw on the same common core skill, strength, or resource).
- Related ratio (the fraction of its revenues attributable to its largest group of somehow related businesses).
- Vertical ratio (the fraction of revenues attributable to the largest group of products, joint-products, and by-products associated with the processing of a raw material through a set of stages).

I think, however, that there is a lack of studies concerning the relationship between strategy and performance in one particular industry and that such studies should therefore be carried out. More specifically, we need to know more about the strategies employed by high-performing companies as compared to those employed by low-performing companies. The main reason for suggesting a focus on a single industry is that generalizations across dispersed industries obviously imply problems in defining and measuring strategy and performance because of the varying premises applying in different industries.

This leads to the exploratory question I posed in an earlier study (Pehrsson, 1995a): Is it possible to identify any strategic state at a certain point in time that is more profitable than another state in the electrotechnical industry? (Chapter 3 of this book presents the strategic states model.) The empirical study dealt with German subsidiaries of Swedish companies manufacturing electrotechnical products and tried to identify the strategic states of high and low performers. (1)

The study suggested that one must carefully consider the advantages to be gained by penetrating just a few selected segments of the German market. A concentration strategy such as this demands a thorough market investigation that would take into account the whole of the German market and then distinguish between specific segments.

Another normative conclusion was that a relatively high share of the invoicing should concern customized products and services, following an adaptation strategy. In order to balance the relatively risky but potentially more profitable customized orders, a company may consider offering a certain amount of pure standardized products. In this question of priorities, access to information about customer preferences is necessary. Considerable variation in preferences will perhaps require that a local product development function must be established.

A final normative conclusion was that the financial objectives for German subsidiaries have to be realistically formulated. A company has to be conscious of the need for financial persistence and has to realize that there are no guarantees that an increase in volume will lead to higher profitability.

Hence, by using cross-sectional data, I found that it was possible to identify a strategic state that was more profitable than another state in a specific industry in a certain market. My study showed that strategic differences also meant performance consequences. This does not, however, imply that the same pattern will be observable at any given time.

NOTE

1. In order to search for the relationship between strategy and financial performance, and to be able to compare the high-performing and the low-performing companies, the empirical study used quantitative cross-sectional data.The data concerned the strategic states and financial performance of West German subsidiaries of Swedish companies manufacturing electrotechnical products. Germany was considered an extensive, highly competitive market for international companies, which would mean that gaining awareness of profitable business strategies employed in that country is of crucial importance.

The strategic states model (see Chapter 3 of this book) was operationalized into six variables, where four represented segment penetration and two applied to the degree of segment adaptation. Based on these independent variables, six hypotheses comparing the strategic states of high-performing and low- performing companies were formulated. Performance at a specific point in time was measured by financial performance as compared to the company's expectations.

The test of the hypotheses showed statistically significant differences between high- and low-performing companies when it comes to the share of the three largest orders in the invoicing, representing segment penetration, and the share of the invoicing for customized orders, representing segment adaptation. That is, two kinds of relationships between strategic states and performance applied to the study sample:

- The three largest orders accounted for a larger share of the invoicing in high- performing companies compared with that of low-performing companies.
- Customized orders accounted for a larger share of the invoicing in high- performing companies compared with that of low-performing companies.

THREE

Models of Business Strategy in Emerging Markets

The framework for business strategy in emerging markets that will be presented in this chapter relies on concepts reflecting a company's process of establishment in an emerging market and the strategy that the company caters for in the market in question (Figure 3.1). The interrelated establishment components form a model for market establishment. Here, the components concern the perceptions of relevant barriers to entering the market, the strategy competence sustaining the efforts, and the chosen strategy for entering the market. The current strategy in the market is captured by the strategic states model. In this model, the dimensions of segment penetration and segment adaptation form

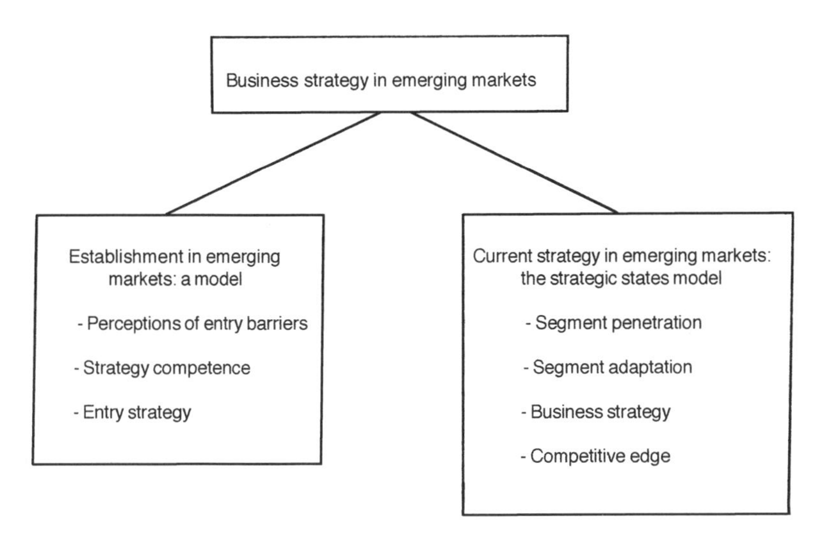

Figure 3.1 Components of models of business strategy in emerging markets.

a matrix in which the company follows a certain business strategy (divergence, concentration, standardization, and adaptation are pure strategies). Each business strategy puts an emphasis on a specific edge, or a combination of edges, for competition in the market.

A MARKET ESTABLISHMENT MODEL

Ellis and Williams (1995) argue that the process by which an organization chooses between different ways of market establishment is dependent on its external and internal context. This is also likely to change over time. At one end of the spectrum, some writers have suggested that the process is, or should be, highly rational and objective, mainly being influenced by the assessment of external factors. Alternatively, other commentators contend that the entry strategy chosen is much more likely to be influenced by internal factors, in particular, by the power elites within the organization and by their vision of the future direction of the organization. In between these extremes is a continuum along which the relative importance of external and internal factors changes.

The theory developed in industrial organization economics for how market structure imposes barriers to entry is one example of the external approach. These barriers imply disadvantages relative to market incumbents for those trying to enter the market in question. Previously established competitors can therefore enjoy higher profitability than they would in the absence of barriers. As first formulated by Bain (1956) and refined by subsequent researchers (for example, Porter, 1980, Yip, 1982), there are distinctive classes of barriers, such as economies of scale, product differentiation, absolute costs, and the capital requirement.

The inherent weakness, according to Ellis and Williams, of relying on the external approach is that this may lead to neglecting the internal context, in particular, the ability to implement externally determined strategy.

Based on the evolutionary strategy perspective, Chang (1996) views entry, and also exit, as sequential search and selection processes. He puts forward propositions regarding the character of the processes, and these are the arguments that I think demonstrate an internal approach to studying establishment processes:

- Motivation for entry or exit.
  What stimulates search and selection at the corporate level? The ultimate goal of the search and selection activities of a firm is to upgrade its knowledge base and thereby to improve performance. Some firms may change their scope preemptively to prepare for changes in external environments that will affect the future profitability of their businesses, such as deregulation or technological innovations in the telecommunications industry.
- Direction of entry or exit.
  In order to reduce the risk of failure in entry, the direction of search activities should be guided by the firm's current knowledge base—hence, a firm should search for markets to enter where it can leverage its existing technology or marketing know-how. This view is consistent with the relatively high performance of related diversification noted in previous research (for example, Rumelt, 1974 and 1982).
- Sequence of entry and exit.
  A firm's movement over time is necessarily path-dependent and evolutionary. Even when a firm wants to enter an industry or market whose knowledge requirements are quite dissimilar to the firm's current knowledge base, the firm can move in that direction by entering several intermediate industries or markets and thereby acquire additional knowledge bases. In this sense, what a firm gains from diversification is an acquisition of additional knowledge.
- Extensive and intensive search.
  There are likely to be limits to how far a firm can leverage its existing knowledge base, causing it to seek a new knowledge base. An intensive search does not need to acquire any

additional knowledge, it only requires exploitation of an existing base. An extensive search, on the other hand, takes place along a new knowledge dimension. This means that extensive search is equal to entry into an unrelated field, and intensive search is equal to entry into a related area.

- Effect of entry and exit on corporate performance.
  A firm will improve its performance as a consequence of search and selection, as individual entry (and exit) decisions collectively contribute to the improvement of the firm's performance.

A major potential weakness of the internal approach is that key decision makers may have a "dated" or inaccurate strategy for the organization:

> In practice, the determination of an organisation's market entry strategy is more likely to be influenced by a mix of external and internal forces. Almost inevitably if one set of factors is over-emphasised, at some stage in the future business performance will suffer and a more eclectic view will need to be taken.
>
> The need to make the right decision is emphasised by the likelihood of there being significant penalties if a wrong decision is made. Not only will there be the investment of management time and financial resources, but/market entry failure may lead to shareholders experiencing a loss of confidence in the management of the company. Equally, for the organisation there is also the opportunity cost, as potentially better opportunities are forgone. Conversely, getting the decision right may help place the organisation on a higher growth path and result in shareholders giving greater support to the incumbent management. (Ellis and Williams, 1995)

Rutihinda (1996) treats both external and internal factors in his broad discussion on entry strategy design. On the external side, he distinguishes between global, domestic, and target market factors. The internal factors are represented by the resources and capabilities of the firm. All these factors have to be considered in the design of entry strategy. Here, decisions have to be taken regarding these issues:

- The type of engagement (green-field or acquisition).
- Ownership structure (sole venture or joint venture).
- The type of value-adding process (own production or contractual arrangements).

- Integration (the degree to which the new venture should be integrated into the company's existing activities).
- Competitive strategy (design of products, pricing, promotion, and distribution policy).

In general, I regard barriers to entry as the major external factor influencing the outline of an entry strategy. The presence of barriers is the ultimate external factor that determines the likelihood of becoming firmly established in a market; however, such barriers are perceived differently from company to company and from one individual to another. This means that, regardless of the "objective" existence of barriers, different perceptions imply varying assessments of business opportunities. Such assessments decide what entry strategy is regarded as appropriate. Conversely, the process of implementing a specific entry strategy might result in changing perceptions as management learns about market conditions throughout the process; therefore, perceptions of entry barriers might change as a result of the implementation process.

In turn, both perceptions of entry barriers and knowledge gained during the process of implementing an entry strategy are strongly related to comprehending the need for internal strategy competence in order to become firmly established in the market. That is, management constantly needs to evaluate to what extent the required competencies are available.

Figure 3.2 presents my market establishment model. In this section of the book, each component of the model will be discussed more thoroughly.

Perceptions of entry barriers

Creating barriers to competition plays an important role in strategic thinking and practice. Much of competitive strategy focuses on how to create and sustain barriers that make it difficult for competitors to succeed (Marsh, 1998). A strategy that promotes the development of proprietary knowledge or brand loyalty or that controls sources of supply, for example, focuses on the market factors that create barriers and the potential to generate financial profits.

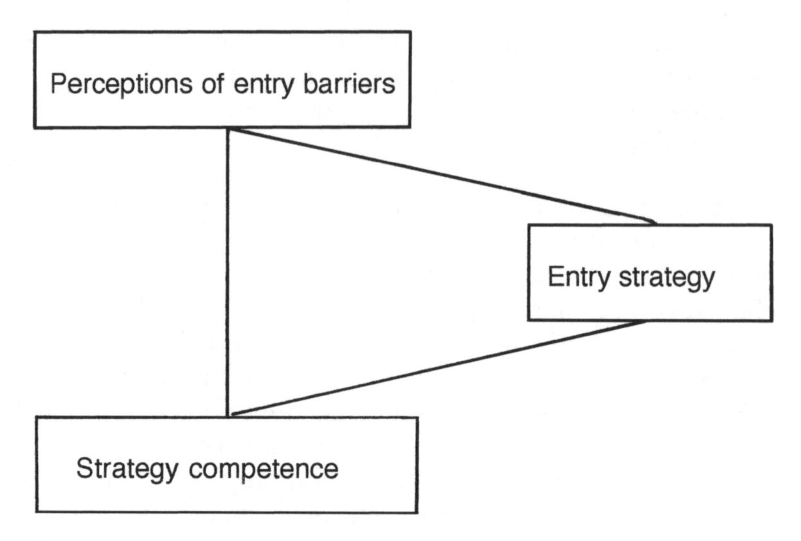

Figure 3.2 The market establishment model.

Porter (1980) suggests that industry competition is driven by four forces external to the industry: the threats from potential competitors and substitute products, and the bargaining power of suppliers and customers. In most cases, potential competitors that want to establish themselves in a market have to face entrance barriers. These barriers may be of different character and importance, depending on the structure of the market, and they may also change over time. Porter discusses seven major types of barriers to entry:

- Scale effects (the new entrant may need enough volume in order to reach low costs).
- Product differentiation (this creates preferences and loyalties among buyers and established sellers).
- Need for capital for investments.
- Switching costs (any customer who wants to switch from one supplier to another faces costs of varying sorts).
- Access to distribution channels (available channels might be too few or controlled by established competitors).

- Important costs independent of scales (for example, access to raw materials, specific innovations, experience, etc.).
- Governmental policies (for example, licences may be required, or interventions may occur).

The importance of an establishment is due to the presence of entry barriers in combination with possible reactions from previously established competitors. Establishments imply that the total supply and capacity will be enhanced and, if the demand does not increase, that the competition will become more intense. As a result, prices may deteriorate and put a strain on the profitability of involved competitors.

Yip (1982) argues that market structure (market growth rate, market concentration, investment intensity, advertising intensity as an indicator of differentiation, and incumbent parent size) alone does not determine the importance of barriers. Entrants also take into account their own specific characteristics and how these allow them to reduce the barriers they face.

In my opinion, present barriers to entry are perceived differently from company to company, and from individual to individual. Although there generally exist some objective facts, perceptions are strongly influenced by subjective interpretations of available information, and this results in different assessments of the importance of barriers, given the uncertainty inherent in judgments.

Milliken (1987) states that managers will generally come up against several types of uncertainty in the course of trying to interpret and respond to the environments of companies. Uncertainty about the state of the environment ("state uncertainty") means that we do not understand the way in which elements in the environment may be changing. An inability to predict the future behavior of a key competitor is a manifestation of state uncertainty. Such environmental uncertainty may imply an inadequate understanding of the interrelationships among elements in the environment. Hence, managers lack information about the nature of the environment. "Effect uncertainty" is defined as an inability to predict the nature of the effect of a future state of the environment or environmental change on the organization. This uncertainty implies a lack of understanding of cause-effect relationships.

Milliken goes on to say that "response uncertainty" implies a lack of knowledge of response options and/or an inability to predict the likely consequences of a response choice. This uncertainty comes closest to the definitions of decision theorists.

I have tried to contribute by measuring state uncertainty in terms of the general manager's total perceived uncertainty and of the uncertainty inherent in judgments of the business climate (Pehrsson, 1985), as well as in judgments of competitors' business strategies (Pehrsson, 1990). The uncertainty in making unequivocal interpretations of gathered and selected information about business climate changes and competition generally dominates the general manager's total uncertainty. A major reason for this is that the business climate and the competition sectors of the environment are less controllable and foreseeable.

Suthcliffe and Zaheer (1998) suggest that competitive uncertainty is the uncertainty arising from the actions of potential or actual competitors. Thus, competitive uncertainty derives from moves or signals by economic actors in current or future competition with the focal firm. The uncertainty engendered by the actions of potential or actual competitors may be deliberate, stemming from strategic motivations. On the other hand, competitive uncertainty may arise innocently from a lack of competitor intelligence or awareness about the prospective actions of competitor firms.

I think that perceptions of entry barriers thus concern perceptions of environmental states and inherent state uncertainty. When difficulty occurs in making unequivocal interpretations of information on entry barriers, or something that may result in barriers, uncertainty arises. Since complete information about the present situation regarding barriers and about expected changes is seldom available in real life, we have to rely on limited information and enjoy only varying degrees of confidence in our assessments. It can also be difficult to select relevant information from the tidal wave of data that inundates the strategic managers of today. This also causes uncertainty.

Some researchers have sought to measure both "objective" and perceived environmental uncertainty and have even tried to examine the congruence between the two measures (for example, Bourgeois III, 1985). Although I admit the existence of objective uncertainty in some instances, my main interest is in perceptions that include some degree of uncertainty.

Strategy competence

Penrose's (1980) view of firm resources implies that a firm's distinctive competence is based on the specialized resources, assets, and skills it possesses and focuses attention on their optimum utilization to build competitive advantage. These tangible and intangible resources are generally under the control of the firm's administrative organ.

Grant (1991) separates firm resources and firm capabilities. Firm resources consist of tangible and intangible resources that may be traded, such as capital equipment, manufacturing systems, skills of individual employees, finances, patents, brand names, and so on. Firm capabilities, on the other hand, refer to a firm's capacity to deploy its resources, such as interactions that govern the production process and management interactions that include monitoring business performance.

As I view it, the likelihood of becoming firmly established in a market depends, to a large extent, on the relevant strategy competence of the company, without any distinction between resources and capabilities. That is, the higher the competence, the better the general likelihood of becoming firmly established in the market. This requires acquisition of knowledge on strategy formulation and implementation, and relevant knowledge about the products, services, and markets involved. Thus, strategy competence is acquired through information- based processes that are firm specific and are developed over time through complex interactions that depend on available capabilities (see Pehrsson, 2000, for details on strategy competence).

Furthermore, I argue that the strategy competence of a certain company is due to the position of its businesses in relation to the original core business of the corporation ("relatedness"). This means that efforts to establish a business that may be considered a part of the original corporate core business hypothetically have the best chance of being successful. The opposite is valid for a business unit that is less related to the core of the corporation.

Besides relatedness, strategy competence includes experience in the market. Familiarity with market conditions clearly strengthens the competence and the likelihood of finding a sustainable position in the market. Thus, increasing knowledge of products, services, and other market conditions gives extended market experience. Such knowledge might be acquired through any sequence of entry or way of searching for knowledge.

The assumption about a positive correlation between strategy competence and the results of the establishment efforts is supported by the findings of Rumelt (1974 and 1982). He found in both studies that corporate profitability differed significantly across groups of firms following different diversification strategies. The highest levels of profitability were exhibited by those having a strategy of diversifying primarily into those areas that drew on some common core skill or resource (related diversification).

The related-core term relies on the definition of a core competence. In general, at least three tests can be applied to identify core competencies of a company (Prahalad and Hamel, 1990). First, core competence provides potential access to a wide variety of markets. Second, a core competence should make a significant contribution to the perceived customer benefits of the end product. Finally, a core competence should be the kind of competence that is difficult for competitors to imitate.

Hamel (1991) suggests that inter-firm competition, as opposed to inter- product competition, is essentially concerned with the acquisition of skills. This means that firm competitiveness is largely a function of the firm's pace, efficiency, and extent of knowledge accumulation.

Strategy formulation and implementation requires analytical, entrepreneurial, and political skills. Lorange and Roos (1993) suggest that analytical skills reflect the ability of the company to carry out relevant strategic analyses and investigations. This demonstrates an overall internal capability to thoroughly gather and process relevant information as a basis for decisions. Further, an entrepreneurial skill aims at reflecting the competence within the organization to bring together qualified and motivated people who can support a particular idea. Such willingness may be determined by how individuals on various organizational levels view themselves relative to the idea. Finally, an ability to deal with a broader set of stakeholder issues requires political skills. For instance, a company may use a stakeholder, such as an outside venture capitalist organization, as a source of professional competence in cooperative venturing, and this may provide additional momentum to various steps in the formation of cooperation arrangements.

Entry strategy

Entry into an emerging market assumes the same character as entry into a foreign market. Whether or not the company in question has been operating in the country previously, the principles of entering into an emerging and, consequently, less familiar market are the same as those of entering a foreign market. In both cases, the company faces less familiar market situations. To a large extent, therefore, the following discussion is based on procedures for entering into foreign markets, and these procedures can easily be transferred into situations valid for entering into emerging markets.

Albaum et al. (1994) argue that entry into foreign markets, initially and on a continuing basis, should be made using methods that are consistent with the company's strategic objectives. A firm becomes committed to international markets when it realizes that it can no longer attain its objectives by selling only domestically. When such a commitment is made, a firm is well on its way to becoming internationalized, even if it is limited only to export operations. It has been observed that exporting may be the best international learning experience, something that takes a firm toward a higher level of sophistication and commitment to other modes of international marketing, such as establishing a manufacturing facility in a foreign market (Root, 1987).

According to Albaum et al., the strategy for entering markets should be viewed as a comprehensive plan, one that sets forth the objectives, resources, and policies that will guide a company's marketing operations over some future time period that is of sufficient length for the company to achieve sustainable growth in the markets.

Very broadly, a market entry strategy can be viewed as a plan for the marketing program to be used for the product/market scope in question. As such, it requires decisions on factors that are relevant to entry into emerging markets in general:

- The objectives and goals in the target market.
- Needed policies and resource allocations.
- The choice of entry modes to penetrate the market.
- The control system to monitor performance in the market.
- A time schedule.

In addition, a marketing plan should include an analysis of the target market and the market environment, a financial analysis, and an evaluation of competitive conditions. Without an entry strategy for a product/market, a company has what essentially amounts to a "sales" approach to foreign markets. This approach has no real commitment to serving markets permanently. Yet, such an approach may be useful for the inexperienced firm and for companies needing to gain experience and greater confidence in their own ability.

A market entry mode is generally an institutional arrangement necessary for the entry of a company's products, technology, and human and financial capital into a market. Albaum et al. suggest seven major alternative modes for entering a market. I view three of these as different ways of organic development, while the other four fall into the category of strategic alliances:

- Organic development (exporting, assembly, manufacturing).
- Strategic alliances (licensing, contract manufacturing, management contracting, joint venture).

In my opinion, market entry through an acquisition also represents a type of organic development, although not originally based on a sole venture. The main reason for this is that an acquisition is one way to carry out an establishment with the ambition of controling the forthcoming process completely. An acquisition entry occurs when an existing competitor in an existing market is acquired by a company not previously established in that market. The acquirer should have the intention of using the acquired company as a base for expansion, not just for holding it as a portfolio investment. Here, the term "acquisition" also includes mergers.

Exporting is perhaps the simplest and easiest way to meet the needs of foreign markets. Export generally has minimal effect on the ordinary operations of the firm, and the risks involved are less than for other alternatives. In indirect export, the manufacturer utilizes the services of various types of independent marketing organizations that are located in the home country. In direct export, the responsibility for international sales activities is in the hands of the producer. If a local company enters a local emerging market, export is not, of course, a relevant mode.

The establishment of assembly facilities represents a cross between exporting and foreign manufacturing. In this entry mode, a manufacturer exports components or parts. At the foreign assembly site these parts, often together with those from other suppliers, are then put together to form the complete product.

The decision to manufacture locally may be forced upon a company because of competitive pressure, market demands, governmental restrictions on imports, or govermental actions that would result in disadvantages for importing. Or the decision may be part of a company's long-range plan to strengthen its international operations. Rarely should a company establish manufacturing facilities as its first international business operation.

Licensing is a type of strategic alliance, and it includes arrangements for the licensee to pay for the use of manufacturing, processing, trademark or name, patent, technical assistance, marketing knowledge, or some other skills provided by the licensor. Licensing is a viable means of developing investment footholds in unknown markets and is a complement to exporting and direct investment in manufacturing facilities. It is often a prelude to a more permanent equity investment.

Contract manufacturing involves contracting for the manufacture or assembly of products by previously established manufacturers, while still retaining the responsibility for marketing. This allows a company to break into markets without making the final commitment of setting up complete manufacturing and selling operations.

In management contracting, a local investor provides the capital for an enterprise, while an international company provides the necessary know-how to manage the local company. This entry mode allows a company to manage another company without equity control or legal responsibility.

The joint venture mode is followed in a foreign market when a non-national company joins with national interests or joins with a company from another foreign country in forming a new company. The central feature of a joint venture is that ownership and control are shared.

There are many ways of classifying strategic alliances. Faulkner and McGee (1997) use this classification:

- Joint ventures (where two companies set up a separate joint venture company).
- A collaboration (where two companies co-operate over a range of activities without forming a separate company).
- A consortium (where a number of companies set up a new consortium company to carry out a major project or activity).

Lorange and Roos (1993) discuss four types of strategic alliances: *ad hoc* pool, project-based joint venture, consortium, and full-blown joint venture. To varying extents, these are characterized by the input of resources by the parents (sufficient for short-term operations or for long-term adaptation), and retrieval of output (to the parents or to be retained in the alliance).

THE STRATEGIC STATES MODEL

One difficulty of the research on business strategy is that two identical strategic settings never occur. This problem has given rise to three views of studying strategy: the situation-specific, the universal and the contingency views (Hambrick and Lei, 1985).

The situation-specific view sees strategy as an artful alignment of environmental opportunities and threats, internal strengths and weaknesses, and managerial values (for example, Andrews, 1987). Proponents of this view are in favor of case research, maintaining that analysts can draw no conclusions about firms' strategies unless they understand the firms' unique positions. Some quantitative research has also demonstrated that strategy generalizations beyond more than one firm can be risky (for example, Hatten et al., 1978).

Hambrick and Lei continue by presenting the universal view of studying strategy. In this view, universal laws of strategy exist and apply to a certain extent in all competitive settings. For example, the Profit Impact of Market Strategies (PIMS) program popularized the "law" of market share, implying its universal applicability in the statement:

> There is no doubt that the relationship can be translated into dynamic strategies for all businesses. (Buzzell et al., 1975, p.102)

Given these laws, any business would be well advised to pursue the strategy of aggressively building up cumulative experience and market share. Such laws imply that there is only one grand type of competitive setting and one universally sound strategy.

The underlying assumption of contingency approaches (for example, Galbraith and Schendel, 1983, Ginsberg and Venkatraman, 1985, Hambrick and Lei, 1985, Hofer, 1975, Miller, 1988, Pehrsson, 1993) is the idea that successful business strategies depend on being able to define an appropriate relationship between variables management controls, such as product development, production and investment decisions, and those variables that are generally outside the direct control of strategic management. The latter non-controllable variables may be defined as environmental variables. These variables consist of market structure, number of competitors, market growth, barriers to entry, and so on.

However, the environment could also be divided into sectors on which management would be generally able to exert some influence. This applies to customers and to the technological level of the market (Pehrsson, 1985), whereas business climate and competition constitute those environmental sectors that are generally less controllable.

Thus, the contingency view means that an optimum business strategy may be formulated, provided certain internal and environmental premises apply to the company. In other words, one assumes a relationship between two independent variables, which influence a dependent variable. The strategy performance of the company would then reflect the degree of its success.

Knowledge and understanding of a company's strategy development concerning composition of the offer, customer categories, and their geographical locations enable the company to position itself in the present and to delineate strategic alternatives for the future. Based on my empirical strategy research and literature studies, I argue that the freedom of action in formulating business strategy is defined by the structure of the state in which the organizational unit of current interest is situated. Thus, in each and every state there exists a certain freedom of action and an optimum business strategy emphasizing a specific combination of competitive edges. This strategy is the optimum way to reach high performance.

The strategic states model that I have developed (see, for example Pehrsson, 1995b and 1996) follows the contingency view on strategy. In the model, the environment of a company is reflected in the dimension of the number of market segments that are being focused on, in which a segment is a limited and measurable part of a larger market. The other dimension is valid for the organization and concerns the adaptation of the offer to the requirements of the customers in the market segments. The offer can be composed of physical products and services in varying combinations.

The degree of the company's segment penetration and adaptation decides its sensitivity to variations in price and other competitive means, as well as the possibilities of adaptation to changes in competition or other environmental fluctuations. Hence, freedom of action varies according to the two dimensions of the model.

This model enables us to describe the strategic development of any company or organizational unit and its current strategic state, as well as the alternatives for the future. Divergence, concentration, standardization and adaptation represent pure strategic alternatives. The model is appropriate even for the classification of competitors and the identification of clusters of business strategies that competing companies follow.

Segment penetration and segment adaptation

In the strategic states model, the breadth of the market penetrated by a company or any organizational unit is specified by the number of segments being penetrated. If the company or unit of current interest follows a divergence strategy along this dimension, and penetrates a larger number of segments than it did previously, its dependence on the single segment decreases. Here, concentration means fewer segments and an increased dependence on each single segment. This is also the case when one or a few segments become dominant in the company's business activities.

A market segment is a limited and measurable part of a larger market. Scandinavia is one geographical segment of the European market, while Sweden could be regarded as one geographical segment of the Scandinavian market. Manufacturers of machines could be viewed as a segment of the manufacturing industry as a whole. In this case it is therefore appropriate to talk about a customer segment that crosses geographical borders.

An effective segmentation procedure results in parts of a market that can be identified and measured. The segments should be large enough in terms of customers' purchasing volumes and in terms of the potential profitability for the companies penetrating the segments. Equally the segments have to be accessible to penetration and defendable against competition.

Once the market segments have been identified, they have to be described. A segment can be described either by the demand characteristics of the customers or by the demographic characteristics of the customers in the segment.

In the strategic states model, an offer is characterized by its degree of segment adaptation. In this dimension, the development can show increasing standardization or adaptation to the requirements of customers in the segments. When it comes to the offer, it can be composed of physical products and services in varying degrees.

Adaptation generally means a significant adjustment of the basic product program from market to market in order to bring it into line with different customer needs and varied market environments. The basic product program may be one followed in the well-known home market or developed after consideration of explicit needs and situations in unknown markets.

I prefer to use the term "customization" when referring to the adaptation of the product program to satisfy fully any customer's requirements. "Localization" is an appropriate term for adaptation of, for instance, the product design to meet the local environmental forces specific to a given geographical market. It usually leads to the development of a diverse collection of product programs with a limited commonality from market to market. Localizing may be necessary and desirable when the conditions of product use and other important factors differ significantly between the various geographical markets.

Thus, dependence on market segments and the degree of adaptation are the dimensions in the strategic states model, and they define a company's sensitivity to variations in price and other competitive measures, as well as its ability to make a quick response to changes in competition or in the environment as a whole. Therefore, the company's freedom of action varies in accordance with the two dimensions in the model. At any given point in time, freedom of action is decided by the structure of the company's strategic state.

The efficiency of a business strategy is measured by performance indicators. Although a broader conceptualization of strategic performance is welcome, I think the financial measures (such as profitability measures) and the operational measures are the most suitable for measuring performance when applying the strategic states model. As regards operational measures, it would be logical to treat such measures as market share, new product introduction, and measures of technological efficiency within the domain of business performance. These measures focus on those key factors of operational success that might lead to high financial performance.

Business strategies: divergence, concentration, standardization, and adaptation

Segment penetration and the degree of adaptation form a two-dimensional space, as shown in Figure 3.3, which illustrates the

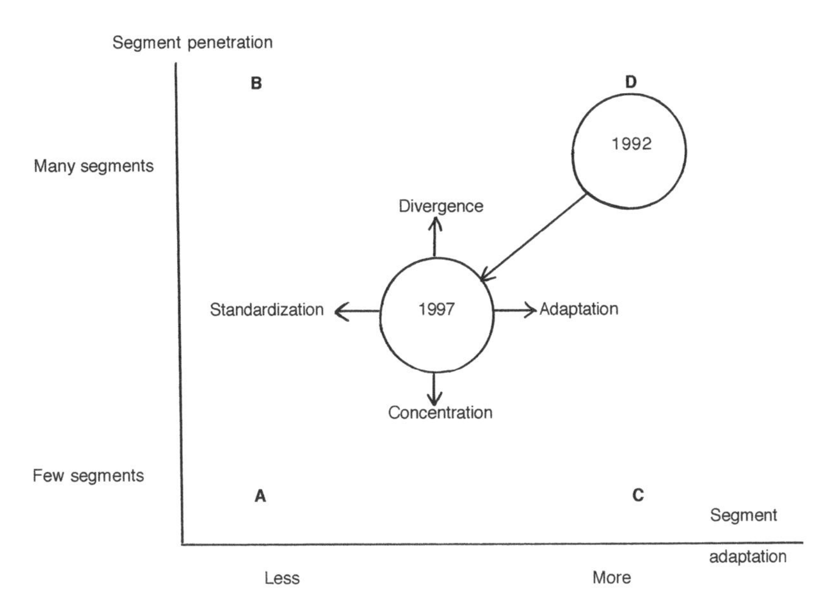

Figure 3.3 The strategic states model: Strategic alternatives of a company based on its state in 1997.

strategic development of a company from 1992 to 1997 and its possible strategic alternatives for the future.

In order to strengthen its financial performance, the company in the example was trying to concentrate on fewer market segments for the five-year period. Earlier, the company had been interested in the whole of Western Europe, whereas in 1997, only Scandinavia was being penetrated. A certain concentration is also valid for the machining industry. In this case, the whole manufacturing industry earlier constituted the market for the strategy that assumed a more divergent character.

The company had been strongly affected by problems with cost calculations when offering relatively well-adapted products. The company therefore carried out a certain standardization of its offer in order to meet the common requirements of the customers in the Scandinavian machining industry. The present standardization strategy is manifested in the effort to be cost efficient and well prepared for price competition.

The example shows that divergence means penetration of a larger number of market segments and, normally, decreased dependence on single segments as well. Concentration, on the other hand, implies an increased dependence. This is also the case if one or just a few segments become dominant.

The degree of adaptation can be adjusted by implementing standardization or by using adaptation strategies. Standardization means making an offer more uniform, while adaptation means that the offer is shaped to suit the unique requirements of customers in various segments.

The concentration strategy could be further implemented by the company in our example. It would then be suitable to segment the machining industry into smaller parts. The main interest should be accorded to companies manufacturing different types of machines and located somewhere in Scandinavia. Companies in attractive segments would be potential customers of the company.

The standardization strategy could be even further intensified. In this strategic alternative, it is necessary to assess carefully the value of the standardized offers when it comes to penetrating various market segments. Perhaps the offer, or parts of it, need to be refined.

Divergence is a business strategy for the future that can be appropriate if the financial performance of the company allows it. In this case new customer groups or geographical regions are sought. New markets for the established products and services will be scrutinized, as well as possible applications of the offer. As the number of segments increases, the company might want to engage distributors to take care of individual customer groups or regions.

As a fourth alternative, a review of the mixture of products and services is conceivable. This is especially valid if the ultimate financial objective has not been reached. A certain adaptation could produce quick results, but it would probably imply a confusion in choosing which competitive edges to emphasize. In order to reduce costs, the company could minimize its own manufacturing capacity and engage subcontractors instead.

Using the dimensions of the model, it is possible to define four extreme states (A-D in Figure 3.3). The type of competition is completely different in these four strategic states. This means that freedom of action and the most efficient business strategies for reaching high performance vary from one state to another.

The relative standardization of products, as well as their concentration by penetrating a limited number of market segments or customers (state A), implies that rationalization and conservative cost control are particularly important. Since customers generally do not perceive any important differences between competing products except for the price, product standardization signifies that price is a decisive competitive factor. High profitability requires that customers within the relatively few market segments demand large volumes.

A company's bargaining power could increase by applying a more divergent business strategy and by offering standardized products to many different market segments or customers without any dominants (state B). Having simple products and numerous customers means, however, that long-term customer relations may be difficult to establish. Such relations normally make it possible to develop and highlight competitive advantages other than price, which means that a company's vulnerability to price competition decreases. Furthermore, divergent strategies and standardized products imply that it is frequently advantageous to cooperate with distributors of various sorts that have wide customer contacts. The ability to handle distribution channels will have a decisive impact on company performance.

A higher degree of adapting products and services to the requirements of customers in single market segments makes it possible to create a more unique offer and to be less dependent on price as a competitive edge. Of course it is still of paramount importance to keep the costs low, even if this may be difficult considering that tailor-made services and subsequent personnel costs are generally more extensive than those for standardized offers.

If the products have been highly adapted to just a few market segments (state C), then high risks are involved. First, segment dependence gives a lower bargaining power. In such a case it may be difficult to argue for a relatively high price, notwithstanding the fact that the price does not presumably decide a customer's choice of supplier. Second, problems with payment or other hardships that the customer must endure might eventually have serious consequences for the company. These would be caused by the dependence and the corresponding restrictions on freedom of action. Third, as the products are presumably adapted to very specific requirements, they cannot be easily available to customers in other market segments.

On the other hand, adaptation makes it possible to build up long-term relationships with customers. In addition to any other benefit, such relationships and the accompanying customer loyalties make it possible to create barriers for competitors threatening with low prices. If costs can be minimized, an effort to create long-term customer relationships can generate high performance in this strategic state.

Extensive adaptation to several market segments, each requiring different product variants, constitutes the fourth and final extreme strategic state (state D). Although the distribution of risks is usually satisfactory and the chances of reaching high performance are good in the long term, this state is quite demanding. Since the variation in customer requirements is extensive, it calls for a large and decentralized organization. At the same time, the high degree of product adaptation makes it difficult to reach scale effects in production and in other functions. High performance is made possible primarily through a search for orders that are significantly similar to each other and, consequently, by trying to exploit the learning effect.

Competitive edges

After the targets have been determined, the competitive mix to be used in serving the market segments must be outlined. This implies an emphasis on a specific edge for competition in the market. Albaum et al. (1994) advocate that, ideally, all elements of the marketing mix (including decisions on products, prices, distribution, and promotion) should be determined simultaneously. In practice, however, some must be determined first, and it is these elements that provide the base upon which the others are determined. For example, in determining the price of products, a manufacturer limits pricing alternatives once a channel has been selected.

In all types of companies, product decisions should be the concern of all management levels. Nevertheless, although top management must lead and control decisions on products, they depend on marketing personnel for information and for planning and implementation of decisions relating to the characteristics of products offered, product lines, branding, and packaging. Albaum et al. continue by saying that the problems are extremely complex for a firm operating in multiple foreign markets. Customers from different countries have varying requirements, and multinational decisions on product characteristics are necessarily quite complex.

First, which breadth of a product range is most appropriate? Should separate products or systems of products be pursued? As it is difficult to realize the implications of a product definition in advance, uncertainty plays an important role in the choice process. Thus, as I have found in an earlier study (Pehrsson, 1993), there is a difference between "product" companies and "systems" companies with regard to perceptions of problems in making product decisions. Generally speaking, volume calculations and related issues are the major problems for "product" companies, especially for those who offer standard products exposed to price competition, while the standardization versus customization issue dominates for "systems" companies. The main reason being that too much standardization means vulnerability to price competition, while extreme customization leads to difficulties in trying to reach scale effects.

Second, decisions have to be made regarding products which are simultaneously offered in more than one market. In a previous study of mine (Pehrsson, 1995b), a pattern of treating product decisions was revealed. The study concerned businesses competing in two separate countries, namely Sweden and Germany. Three ways of handling product decisons valid for the two markets emerged:

- Emphasis on similarities, regarding both user characteristics and product design, across national boundaries.
- Emphasis on pure product standardization, without specific attention to user requirements.
- Emphasis on well-defined product adaptation, covering both customization and attention to other local demands.

Broadly speaking, pricing decisions include setting the initial price as well as changing the established price of products from time to time. Frequently, price decisions must be made for different classes of purchasers, such as consumers or industrial users, wholesalers or distributors, licensees or one's own subsidiaries or joint ventures. The issue of differential pricing is important, and decisions must be made on whether the price to customers in one market should be the same, lower, or higher than to those in other markets.

The structure of distribution for reaching any market segment includes all of the intermediary agencies or institutions that are in use at any given time, their capacities and capabilities, and their geographical coverage. Albaum et al. say that, in developing its entry mode, a company must plan for the flow of transactions and for the flow of physical product through the distribution structure. Many specific types of organizations may be involved in performing the transactions and physical flows in a given channel of distribution.

Communication is a major part of marketing activities. Multimarket communication is basically a cross-cultural communication; that is, communication between a person in one culture and a person in another culture. It is, however, possible for certain segments of markets in, for example, different nations to be culturally similar. Essentially, the promotion decisions can be reduced to: What messages? What communications media? How much effort or money to spend?

PART II

Telecom Operators in the UK and Sweden: Industry Overview

INTRODUCTION

As the UK was a pioneer in the comprehensive international deregulation process within telecommunications that started in the late 1970s, it is of great interest to study the development of strategies of the operators in this market. The deregulation started early in Sweden also and it is interesting to describe and compare the situations in these two markets.

Thus, the purpose of the exploratory cross-sectional study presented in Chapters 4 through 6 is to describe and examine strategies of network operators in the UK and Sweden. More specifically, what is the outline of establishments and initial strategic states? The operators in question are those who held public telecommunications operator's licences for stationary or mobile telephony in 1996. Licences might be valid for local, national or international telecommunications. This means that operators offering telephony services without possessing network licences are not included in the study.

Chapter 4 consists of operationalizations of concepts in the framework for describing strategies in emerging markets (outlined in Chapter 3) and the description of the methodology of the study. Chapter 5 comprises information on operators' establishment in the UK market and their strategic states, while Chapter 6 presents the corresponding information for the Swedish market.

The main results of the study are summarized in the following:

- Any company that has a desire to enter the industry will have to face the need for capital as regards the arrangement of infrastructure. No matter whether the operator tries to build its own network or whether it rents capacity, the capital demand is extensive. Nevertheless, alternatives to the costly traditional technology based on copper cables are emerging in the form of fiber optics and radio technology. In the UK market, there are several companies with telecom operations that do not belong to the original corporate core businesses. For example, electricity companies and cable TV companies have entered the industry by exploiting their networks that were originally established for other purposes. In fact, most foreign entry started in the cable TV industry.

- Operators with most market experience in both countries, including the dominant British Telecom (BT) in the UK and Telia in Sweden, follow a resource- demanding strategy of divergence combined with a high degree of customization. These operators are being challenged by late entrants.
- All of the UK and Swedish operators that are situated in a state of concentration to company customers and the offering of broader communications services, partly including telephony mainly using leased lines, entered the markets late. Concentration and adaptation is, thus, a common strategy combination for entering the markets. This might be viewed as a basis for further strategy developments, and ambitions to be established in other strategic states.
- As regards competitive edges, there is a pressure for low prices as a means of penetrating telephony customers in both countries. Price competition is particularly intensive in international telephony, characterized by standardized services that are difficult to differentiate and by numerous competitors. Further, efficient distribution systems are a crucial establishment condition for operators, such as mobile telephony operators, that are penetrating both companies and private consumers.

If we now turn to the processes of deregulation, which constitute the background of the study, it is pertinent to start the description in the late 1970s. Primarily because of changing political attitudes at that time, a comprehensive process of deregulation was initiated in the UK. In 1981, telecommunications activities were separated from the other activities of the post and telecommunications administration, making way for the establishment of BT. The areas that had to do with apparatus and equipment that were to be connected to the BT network were also opened to competition by the British government in 1981.

The British Telecommunications Law of 1984 made possible a privatization of BT, but before this happened, the Office of

Telecommunications (Oftel) was established as a regulator. Oftel grants operator's licences and decides accompanying terms. This authority is also engaged in the standardization of technology and approves technologies, products, and systems for the market. The BT licence of 1984 included more than 50 stringent conditions that would control its market dominance and would ensure that it did not abuse its position in a way that would run contrary to the public interest. Among the conditions were the following:

- An obligation to serve all customers without discrimination or preference and to provide basic telephony services and any other services where the demand is reasonable. BT also faces social obligations, such as emergency services.
- An obligation to assist its competitors in pricing announcement rules, interconnection rules and non-discrimination rules. Interconnection rules compel BT to open up the public networks so that competitors may reach all customers.
- A prohibition of engagement in cable TV activities.

In 1982, a national operator's licence for public telecommunications, using stationary networks, was granted to the Mercury consortium. In 1991, the BT and Mercury duopoly was deregulated even further and, regional licences were granted to cable companies and telecommunications operators. Thus, cable TV operators were licenced to offer not only television but also phone services across a local network.The UK market became accessible to foreign operators that originated from countries with deregulated telecommunications markets. The next step in the deregulation took place in 1996, and international facilities-based services were liberalized.

Although there were a large number of operators across all segments of the market, in 1995 BT was the only operator with a national network that was active in all market segments. In that year, BT still had over 95 percent of all exchange lines (Oftel Statement, 1995).

Table 4.1 presents statistics concerning shares of the number of international and domestic long-distance telephone calls through

Table 4.1 Shares of the number of telephone calls through stationary networks in the UK in December, 1996 (source: Oftel Annual Report, 1997).

	International calls		*Domestic long-distance calls*	
	Companies	*Private consumers*	*Companies*	*Private consumers*
BT	53%	81%	73%	90%
Mercury	23%	9%	19%	6%
Cable TV operators	2%	8%	2%	4%
Others[1]	22%	2%	6%	

[1]"International Simple Resale" operators (i.e. specialized middlemen between users and network operators).

stationary networks in the UK at the end of 1996. As regards local calls, BT had around 95 percent of the market.

During 1996, cable TV operators won around 60,000 private consumers each month from the customer base of BT. The total number of consumers increased by 65 percent during that year, while the number of companies using cable telephony increased by 78 percent. At the beginning of 1997, there were more than 100 cable TV operators offering telephony services in the UK (Competition in UK Telephony, 1997).

In mobile telephony, the Cellnet operator, with BT as the majority owner, received its licence in 1983. This was also the case for Vodafone, while a Mercury subsidiary started to trade the One2One brand name shortly after that. Orange was the fourth operator to join the British mobile telephony market.

At the beginning of 1997, around 6.9 million subscribers used the services of these operators, with the majority subscribing for GSM services. In January, 1997, Vodafone had 42 percent of the calls volume, Cellnet 38 percent, One2One 12 percent, and Orange 8 percent (Utility Week, 1997).

If we now turn to Sweden, we can conclude that, unlike many other European countries, there has never been a legal monopoly in this country for the establishment of telecommunications networks or the offering of services. However, Televerket (the Swedish public telecommunications administration) historically had a monopoly-like hold on many sectors of the industry. Different monopolies on access to the public network underscored the market power of Televerket.

The Swedish situation has changed step by step. In 1980, the markets for answering machines and telefax equipment were opened up. The monopoly on connecting telephone terminals to the public network was revoked in 1985, and the corresponding monopoly for office exchanges came to an end in 1988. Moreover, there are no regulations protecting Swedish interests or restricting foreign operators from establishing themselves in this country.

The Swedish market is often considered one of the most deregulated markets. This is one example of such a view:

> MFS regards Sweden as the most deregulated market in Europe. With a national carrier that was not a typical PTT service provider and which never had a legally mandated monopoly, Sweden may well be the most liberal and regulation-free market in the world today. In fact, competition in wireless and other services began there even before Sweden had a general licensing regime and independent regulator. As a result, the market is fully open to competition in all domestic and international market segments. (MFS Communications Company, Annual Report 1995)

As is the case for BT, Televerket has gone through organizational changes in order to be able to compete more effectively. A major step took place in 1993 that involved the conversion of the operator into a group with a parent company, called Telia, and subsidiaries. Telia is partly owned by the state as a result of the privatization process and the sale of shares in the company in 2000. Like BT, Telia has certain obligations. For instance, Telia has to provide telephony services throughout the whole country, without discrimination.

Along with the creation of Telia, a specific telecommunications law was established to make possible effective competition, among other things. The National Post and Telecom Agency is the Swedish control authority on the law. This authority took over those regulation activities that Televerket was previously responsible for.

Table 4.2 Shares of revenues in the Swedish telephony market using stationary networks (source: National Post and Telecom Agency, 1997).

	International calls	*Long-distance calls*	*Regional calls*	*Local calls*
Telia	73%	88%	93%	98%
Tele2	22%	11%	6%	2%
Others	5%	1%	1%	–

Permission of different types and accompanying terms, as well as interconnection rules, are its main areas of responsibility.

The telecommunications law was revised in 1997. This meant, among other things, that the agreement between the Swedish state and Telia was not continued; instead, general rules were formed in order to secure state interests.

Telia has large shares of all segments of the Swedish telecommunications services market, while its shares of segments for various types of equipment are smaller. Table 4.2 shows market shares in stationary telephony.

Regarding mobile telephony in Sweden, Table 4.3 presents the number of subscriptions for the competing operators and their shares of the total market volume.

Table 4.3 Number of mobile telephony subscriptions and percentage of market shares in the Swedish market (source: Annual Reports 1995 and 1996).

	Year end, 1995		*Year end, 1996*	
	Number	*Share*	*Number*	*Share*
Telia (the NMT system)	967,000	48%	921,000	37%
Telia (the GSM system)	463,000	23%	824,000	33%
Comviq GSM	422,000	21%	466,000	19%
Europolitan	148,000	7%	281,000	11%

FOUR
Design of an Industry Study

Following the framework for strategies in emerging markets, this chapter presents the outline of a cross-sectional study of strategies of telecom operators in the UK and Swedish markets. Operationalizations of the constructs are followed by the methodology of the study.

MEASUREMENTS OF ESTABLISHMENT AND STRATEGIC STATES

Figure 4.1 summarizes the measurements of establishment and strategic states in the study of strategies of telecom operators in the

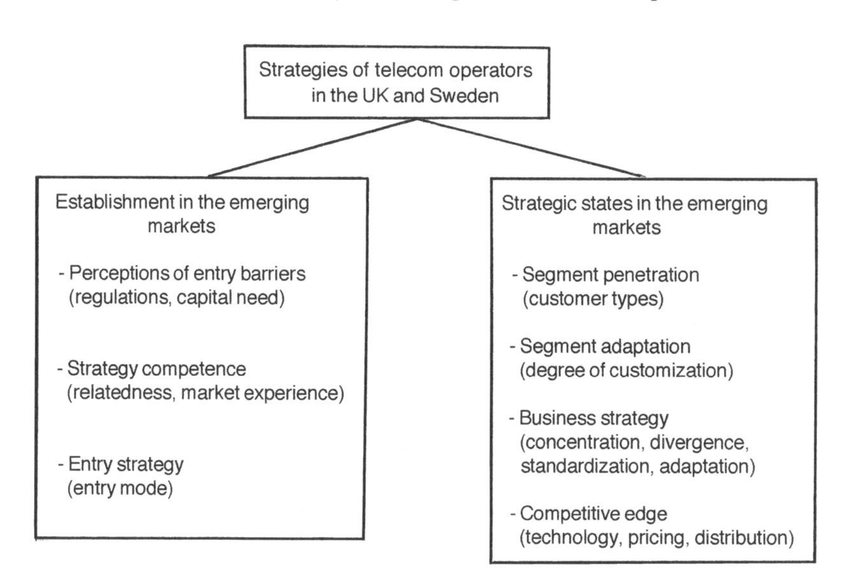

Figure 4.1 Measurements of establishments and strategic states in the study of telecom operators in the emerging markets in the UK and Sweden.

UK and Sweden. Figure 4.2 summarizes the structure of the information in the study. Each of the information alternatives are rooted in the general framework presented in Chapter 3.

Measurements	Information content
Establishment	
- Perceptions of entry barriers	
* Regulations	Perceptions of deregulation
* Capital need	Infrastructure investments
- Strategy competence	
* Relatedness	Relation to the original corporate core business
* Market experience	Early entrance, late entrance
- Entry strategy	
* Entry mode	Organic development (sole venture, acquisition), strategic alliances
Strategic states	
- Segment penetration	
* Customer types	Companies, consumers
- Segment adaptation	
* Degree of customization	Telephony, communications services
- Business strategy	
* Concentration, divergence, standardization, adaptation	Strategy clusters: concentration/standardization, divergence standardization, concentration/adaptation, divergence/adaptation
- Competitive edge	
* Technology	Stationary networks, radio-based networks
* Pricing	Pricing policy
* Distribution	Distribution structure

Figure 4.2 Information alternatives for measurements in the industry study.

Establishment in the emerging markets

The present study of network operators focuses on two types of barriers to entry, as outlined in the framework: governmental regulations and the need for capital (Figure 4.1). One main reason for omitting the other five types of entry barriers is that scale effects and costs independent of scales are very difficult to measure, as this requires detailed company information. Product differentiation, switching costs, and access to distribution channels involve assessments of relationships among operators and their customers, which is not a major topic for this study.

Regulation aspects are captured by information on relevant perceptions, including views on the roles of the regulators, Oftel in the UK and The National Post and Telecom Agency in Sweden. Capital needs are discussed in terms of investments in the construction of infrastructures (Figure 4.2).

In this study, the theoretical construct of strategy competence is measured by relatedness and market experience. Information on the pertinent relationships between telecom operations and the original corporate core businesses indicate the relatedness aspect of competence. Market experience is reflected by the point in time at which the market in question was entered. Operators entering the UK and Swedish markets before 1991 are referred to as "early entrants," while those entering after that year are called "late entrants". The reason for choosing 1991 as the breaking point is that the duopoly of operators, consisting of BT and Mercury in the UK, was deregulated in that year and regional licences were granted to other operators and cable companies. The number of competing operators in Sweden also increased in 1991.

Entry modes represent the broader term of entry strategies, and information on organic developments of sole ventures and of acquisitions and on strategic alliances reflects modes of entering the emerging markets.

Strategic states in the emerging markets

In the application of the strategic states model, classifications of competing operators are made according to the initial state appropriate for each operator. Thus, clusters of business strategies that competing companies follow are identified. These are the strategy clusters available: concentration/standardization, divergence/standardization, concentration/adaptation, and divergence/adaptation.

The segment penetration in 1995 is expressed by the customer types that each operator caters for (companies, or companies and private consumers). The degree of segment adaptation in 1995 is manifested by the type of end-products (telephony, or broader communications services, including telephony), representing various degrees of customization.

This model application is similar to the application of the strategic group concept, which results in maps of competing companies. For mapping groups of competitors, Porter (1980) suggests a number of useful principles. First, the best variables to employ as dimensions are those that determine the key mobility barriers in the industry. Mobility barriers deter the movement of firms from one strategic position to another. Different strategic groups represent different levels of barriers, which provide some firms with persistent advantages over others. Second, in mapping groups, it is important to select as axes variables that do not move together too much. For example, if all firms with high product differentiation also have broad product lines, then the two variables should not both serve as axes on the map. Rather, variables should be selected that reflect the diversity of strategic combinations in the industry. A third principle of Porter is that an industry can be mapped more than once, using combinations of strategic dimensions.

In a survey study of electrotechnical and telecommunications companies (Pehrsson, 1990) I sought to throw fresh light on the concept of the strategic group, and I also paid attention to perceived uncertainty in assessment of competitors. After factor analyses, four key dimensions were proposed for grouping competitors according to their business strategies. Thus, competitors could be classified according to the breadth of their market scope and according to their degree of adaptation to customers. Product quality and pricing can also be used as dimensions in identifying strategic groups.

The strategic dimensions can be used for several maps of competitors. In applying any pair of dimensions, however, the analyst should check that the two values of the pair do not move too closely in the same direction. When it comes to mobility barriers, product quality or prices should be the easiest dimensions in which the individual firms can make adjustments. Changes in market scope or in the degree of customization, on the other hand, are a more long-term matter and not very easy to realise. Although mobility barriers must be considered in each specific case, market scope and customization generally determine the main barriers.

Since the penetration of a widespread market is a resource-demanding affair, distributors often play a dominant role. Hence, control of marketing channels is frequently a decisive mobility barrier to strategic groups operating in such markets. Another example of mobility barriers is the kind of intensive, long-term customer relations that bring advantages to strategic groups characterized by a high level of individually adapted products.

In the strategic states model, the market scope of a company is specified by a "number of segments that are being penetrated" (segment penetration), while customization is an operationalization of segment adaptation. Thus, one conclusion from the discussion above is that the strategic states model is appropriate even for the classification of competitors and for the identification of strategic groups. Hence, a company or any other organizational unit has the ability to group competitors, search for a defendable position and develop its own business strategy following the strategic states model.

Competitive edges are the final component of the concept of strategic states. For example, a competitive edge might be of technical character, or based on efficient pricing or distribution. The study focuses on information emphasizing stationary networks or radio networks, pricing policy (low prices in general, or differentiated pricing according to the customer types catered for are ways of handling the pricing issue), and distribution structure.

Transmission technologies are often referred to as technology edges in the telecommunications industry. In principle, three technologies exist for the transmission of communications signals: traditional copper cables, fiber optic cables, and radio transmission.

The national networks of the public telecommunications administrations, which have been built up over decades, primarily consist of copper cables. The fiber optic technology provides, however, a higher cable capacity at a lower cost, which means that network costs tend to be less dependent on geographical distances. (This is particularly obvious in international traffic, where competition among network operators is intense.)

The development of the cellular radio for accessing public telephony networks introduced mobility, and the subscriber was no longer tied to any particular location to make and receive calls. During the last decades a number of analog and digital mobile telephony systems have been introduced, which follow different standards.

If an operator wants to offer international telecommunications to large customers, it is normally enough to supply a local loop that connects the customer to the long-distance and international lines. The operator might own these lines itself or rent them from another network operator.

High costs for cables and other equipment imply that households and small companies are frequently not profitable when it comes to direct connections; therefore there is a need for a so-called local access network. The traditional technology for local access is based on copper cables. As companies such as BT and Telia possess established local networks, new operators are, to a large extent, dependent on interconnection and use of these existing local networks.

Alternatives exist, however, for local access to households and small companies. Cable TV companies may exploit their networks in order to offer telecommunications services as complements to ordinary services. Radio technology is another option in local access networks.

METHODOLOGY OF THE STUDY

In March 1996, Oftel provided a list of licence holders for public telecommunications networks for stationary and mobile telephony. The list contained 24 operators in the UK. These companies then received a letter asking for an annual report and relevant information presenting the company and its strategy, such as information on products, services, and target markets. After a reminder, 15 operators had replied. This gives a 68 percent response frequency for the UK part of the study. The following companies replied:

- Atlantic Telecommunications Ltd
- British Telecommunications plc
- COLT-City of London Telecommunications Ltd
- Kingston Communications (Hull) Ltd
- Mercury Communications Ltd
- Mercury Personal Communications
- MFS Communications
- Orange Personal Communications Services
- Racal Network Services Ltd
- Scottish Hydro-Electric plc
- Scottish Power Telecommunications Ltd
- Telstra UK Ltd
- Videotron Ltd
- Vodafone Ltd
- Worldcom International Inc.

Two UK operators were not able to participate in the study because the companies (Liberty Communications Ltd and South Western Electricity plc) were not operational at the time. In addition, the following UK operators did not supply any information: AT&T (UK) Ltd, Energis Communications Ltd, Ionica L3 Ltd, Norweb Communications, Sprint Holding (UK) Ltd, Cellnet, and Torch Communications Ltd.

In August 1996, the National Post and Telecom Agency provided a list of licence holders in Sweden. This list contained 12 operators of stationary or mobile telephony. In accordance with the limitations of the study, companies offering telephony services without possessing network licences were not included in the sample. Following the UK procedure of asking for information, ten companies supplied the requested information. This means that the Swedish part of the study gave an 83 percent response. Nordiska Tele8 and AB Stokab did not provide any information, however, I was able to use information on Nordiska Tele8 that was available in a newspaper interview. Altogether, this resulted in 11 companies with licences in Sweden:

- Telia AB
- Tele2 AB
- Telenordia AB
- France Telecom (Global One)
- Dotcom Data&Telecommunications AB
- MFS Communications AB
- Telecom Finland AB
- Europolitan AB
- Comviq GSM AB
- Banverket
- Nordiska Tele8 AB

The participating companies provided annual reports, internal documents, and newsletters. I then used these information sources, added to information from newspaper interviews and other sources, to compile strategic profiles of each company. These profiles constituted the basis for exploring patterns.

Thus, under the general category of qualitative source materials, a variety of information was gathered concerning strategic profiles. Efforts were made to reduce the information to an accessible form. The first step meant extracting direct and indirect quotations made by responsible managers of the companies. These quotations were sorted into those which dealt with issues of a strategic character in line with the framework of the study, and into those which were not highlighting such major issues. Other information was handled similarly. Different sources were checked against each other for the purpose of verification.

As all written materials were not gathered, a formal content analysis was not possible. However, the available information was enough to fulfill the objective of the collection process: to compile strategic profiles of the companies.

The companies had the opportunity to comment on the compiled information. The external validity was further checked through personal interviews.

FIVE

Telecom Operators in the UK

The description of network operators' strategies in the UK market follows the measurements of establishments and strategic states presented in Chapter 4. Establishment processes are exemplified by company illustrations in the first section, while the second section presents clusters of operators following certain strategy combinations and competitive edges.

ESTABLISHMENT OF TELECOM OPERATORS IN THE UK

First, the establishment concept is composed of a component pertaining to perceptions of two types of entry barriers (governmental regulations and the capital needed for infrastructure investments). The strategy competence component of establishment concerns relatedness between businesses, and market experience, while entry strategies are reflected in entry modes.

Perceptions of entry barriers

A debate is going on in the UK regarding the role of the regulator, Oftel. For example, Mercury's Chief Executive Officer demands increased powers for Oftel. This is his perception of governmental regulations:

> One obstacle in our path continues to be over-complex and increasingly unworkable regulation. This now threatens to impede the growth of competition, with customers as the losers. We are fast approaching regulatory failure, which only fundamental reform can resolve. Our recommendations for this reform require increased powers for the Director General of Oftel to curb anti-competitive behaviour and the removal of artificial barriers to competition. (Chief Executive Officer Peter Howell-Davies, Mercury Communications, Annual Report, 1995)

On the other hand, Oftel is to a large extent concerned with BT's behavior in the deregulation process. This is an example of Oftel's view:

It is important that BT should recognise that its own behaviour will be a major determinant of whether the UK market can be considered one of effective competition. At the moment BT too often appears reluctant or unwilling to address with Oftel and the industry the issues of clear regulatory concern. It needs to have a more positive approach to improving the competitive framework in the UK. Further,

- It needs to have in place convincing and effective internal measures to ensure the compliance of the whole organisation with regulatory and competition rules.
- It should be able to point to an internally independent compliance function and management systems if it is to convince Oftel that it is serious about competing fairly within the rules of a less interventionist system.
- BT also must be able to demonstrate to Oftel that it has in place the management information systems necessary both for Oftel and its own management to know whether or not it is complying with its licence obligations. Recent cases have demonstrated the inadequacies of BT's systems, although the introduction of accounting separation and the detailed reporting requirements it includes should greatly improve this. (Oftel Statement, 1995)

Regulation has a significant impact on BT, both in the UK and overseas. BT is active in lobbying for the liberalization of the telecommunications market in continental Europe and for more regulatory consistency around the world:

In Europe, it is encouraging that the European Union is demanding the full liberalisation of the telecommunications market by 1 January 1998. But the industry in many European countries is still controlled by state-owned monopolies and the market will only open up if the pressure for liberalisation is maintained. Mainland Europe is now our home market and we are establishing a significant presence in many European countries: in Germany through our alliance with VIAG; in Italy through our joint-venture with the Banca Nazionale del Lavoro; in Scandinavia through relationships with the Norwegian and Danish operators. This comes on top of the deal with the Banco Santander in Spain, which we announced last year. (Chairman Sir Iain Vallance, British Telecommunications, Annual Report, 1995)

In the UK, BT views the progress towards liberalization as uncertain and the regulatory environment as hostile and unpredictable:

In the UK, we face significant competition from the cable companies and other licenced operators. Competition is flourishing in the local, long-distance, international and mobile markets. But, provided that it is fair and equitable competition, we welcome it as in the best interest of customers and the industry as whole.

When I wrote to you this time last year, I suggested that in the UK we suffered from "the paradox of greater competition accompanied by greater regulation." This year, I believe that the relationship with our regulator, Oftel, has undergone a step change. Although BT accepted Oftel's proposal for accounting separation, which requires us to publish more detailed cost and pricing information, we were not able to accept the regulator's proposals on number portability. Consequently, Oftel announced that it would be referring the issue to the Monopolies and Mergers Commission (MCC). This is the first time BT has experienced such a referral, but we are convinced that we are right to reject Oftel's proposals.

Number portability refers to the customers' right to keep their existing phone number even if they change phone suppliers. I wish to make clear that BT supports this principle, but we disagree with Oftel about how the costs should be allocated.

The MCC reference is just one example of how Oftel requires us, in effect, to subsidise competitors, many of whom are large, rich American companies which, ironically, enjoy monopoly or near-monopoly status in their home markets.

BT's trading prospects are sound but susceptible to the effects of a hostile and unpredictable regulatory environment in the UK and the uncertain shape of regulation overseas.

Outside the UK, BT is already recognised as a global leader. That leadership will not be maintained, however, if BT continues to be unduly handicapped in its home market. The Government and regulator need to balance their understandable drive to promote competition in the UK with a proper regard for the UK's competitiveness in the global market.

The time, we believe, is right to review whether the incentives to competitors remain appropriate and whether the constraints on BT can begin to be relaxed and free competition allowed to flourish. (Chairman Sir Iain Vallance, British Telecommunications, Annual Report, 1995)

Need for capital in order to become established in the market is the second entry barrier on which this study focuses. Examples of capital needed for infrastructure investments will be given in the following:

- In 1994/95, BT had 13,893 million pounds in turnover and 2,671 million pounds in capital expenditure, which was an increase of 23 percent from the previous year. Over 270 million pounds were spent on a range of research projects.
- In 1994, Mercury announced an investment of 200 million pounds in intelligent networking, the first part of a worldwide billion-pound initiative by parent company Cable&Wireless.
- Mercury Personal Communications had invested around 660 million pounds by December 1996 in building and expanding

the service area in London and the South of England, the Midlands, and North West Regions.

- With the financial backing of its parent company Scottish Power, Scottish Telecom has invested in a multi-million-pound fibre optic network spanning Scotland's central belt.
- Around 40,000 kilometers of fiber optic cable have been laid out by British cable TV operators. These may offer both telephone and television services within licenced areas. The civil engineering investment in the network has been funded by private investors.

Besides the building of networks, finance leasing is an option exemplified by the Atlantic case. Thus, the Atlantic Telecommunications Company is expected to require funds from the Scottish parent company, CaledonianMedia Communications, in order to develop the business beyond the concept stage. Initially, however, much of the telephony switching and network expenditure will be purchased with finance leases giving the group an opportunity to develop the concept. Funds will be required, however, before they launch a commercial service.

Another option is, of course, to concentrate investments in certain areas. For example, the corporate strategy of Kingston Communications involves the growing of businesses around its core telecommunications know-how, while maintaining an investment program in its own public network. The technical focus addresses network management, network infrastructure, and services. Kingston believes that the quality of network management is decisive in the delivery of customer-oriented services.

The Vodafone Group concentrates its investments in foreign markets. In the UK in 1995, the Group had a turnover of 1,153 million pounds, and its operating profit was 353 million pounds. Building on the results in the UK, the Group is expanding into foreign markets. The focus is on Western Europe and the Pacific Rim, where it has subsidiary, associate, and equity investments. In 1995, 68 percent of the Group's capital expenditure and investment was incurred overseas. Many of the overseas businesses are still in the start-up phase and are showing a loss, but they are expected to make a profit.

For other companies, investments are major explanations for negative financial results. One such company is MFS Communications. This company group operates through subsidiaries in telecommunications services and network systems integration.

In 1994, telecommunications services had a revenue of USD 228,707 million and an operating loss of USD 131,216 million, while network systems integration reported a revenue of USD 155,869 million and an operating loss of USD 4,913 million.

> Given the rapid pace of MFS' new investment and new service deployment, we are often asked when we expect our company to begin generating operating cash and to become profitable. We continue to believe that 1995 is expected to be the year in which our operating cash flow losses will bottom and begin a trend toward positive results. (Chairman and CEO James Q. Crowe, MFS Communications Company, Annual Report, 1995)

Strategy competence

The likelihood of becoming firmly established in a market is, hypothetically, based on the strategy competence of the company in question. As the framework of this study shows, the strategy competence of a corporation is composed of the closeness between its businesses and the original corporate core business and its experience in the market.

Figure 5.1 shows the classification of UK operators based on relatedness and market experience. Relatedness is covered by information on the closeness between the operation of telecom networks and the original corporate core business, while information on early or late entrance reflects experience in the UK market.

BT, with its former monopoly, was one of the earliest entrants into the UK telecom market. The company has its origin in the operation of telecom networks and has grown around this core. The mission of the company involves the provision of telecommunications and information products and services, and developing and exploiting networks at home and overseas.

Relatedness	Market experience	
	Early entrants	Late entrants
Telecom operations are part of the original corporate core business	British Telecom Kingston Mercury Mercury PC Vodafone	COLT MFS Scottish Telecom Telstra Worldcom
Telecom operations are not part of the original corporate core business		Atlantic Orange Racal Scottish Hydro-Electric Videotron

Figure 5.1 Classification of UK operators according to relatedness and market experience.

Although BT remains the UK's largest supplier in the industry, it continues to develop its global presence. Initially, BT's strategy was to focus on providing network-based services to multinational customers and their extended enterprises and to major nationals in Europe, North America and the Asia-Pacific region. In 1995 BT had around 30 overseas offices and employed 2,600 people outside the UK.

Kingston Communications (Hull) was originally established in the early 1890s, when the city corporation of Kingston-upon-Hull made plans for a municipal telephone system to compete with the Post Office and the privately owned National Telephone Company's (NTC) networks. In 1904 the first exchange was opened, competing for customers with the Post Office and NTC. The Postmaster General then secured a UK monopoly of telephone services, buying out the NTC and many local authority-owned services. Hull's bid for a new licence was conditional on the purchase of the NTC network in the city. The council approved the purchase, and the sole municipally owned telephone company survived.

In 1995, the Kingston Communications group operated a stationary telephone network serving 170,000 customers. The company remains municipally owned and has undertaken an expansion program based around its core telecommunications know- how. The diversification of recent years has been replaced by consolidation, taking a long-term view on its performance.

Part of the consolidation has involved the formation of a joint venture with Yorkshire Electricity, called Torch Telecom. Addressing the business market, Torch Telecom has launched a range of services in the Yorkshire and Humberside region.

The Kingston group also includes business units for software development, sales and service of telecommunications to business organizations, information services, testing laboratories, and satellite-based communications.

Mercury Communications (later a part of Cable&Wireless Communications) is the third early entrant that has telecom networks operations as a part of the corporate core business. The company was formed in 1981 as a partnership between British Petroleum, Cable&Wireless, and Barclays Merchant Bank in order to compete with BT. However, by the end of 1984, Cable&Wireless became the sole owner by acquiring all the shares.

Mercury offers a full range of telephone and data transmission services, both nationally and worldwide. The company is divided into six business units that are supported by customer service, network, and product development: International Business Services, Corporate Business Services, General Business Services, Business Enterprise Services, Home Business Services, and Partner Services.

In 1991, cable TV operators in the UK were licenced to offer not only television but also phone services across a local network. The Cable&Wireless response to this was to buy 20 percent of the holding company for Bell Canada Enterprises, who operate a UK cable television and telecom franchise. In return, Bell Canada bought 20 percent of Mercury shares. This strategic alliance offered mutual benefits: Mercury was able to bypass the use of BT lines for local calls, and Bell Canada was able to offer its customers cheaper long-distance and international calls.

The UK-based Vodafone Group aims to be a world leader in mobile telecommunications, although the UK market in 1995 totally dominated the turnover. The businesses comprise a cellular radio network operation and service provision, value-added network services, radio paging, packet radio, and the design and manufacture of subscriber equipment and infrastructure. Launched in 1985, the Vodafone analog cellular network had over 2 million subscribers in 1995, both in business and in consumer markets. In 1993, Vodafone launched its digital network in Central London, which conforms to the GSM standard. In 1995, this network covered around 95 percent of the UK population and more than 3,500 radio base stations were deployed.

City of London Telecommunications (COLT), MFS Communications, and Worldcom belong to telecommunications groups with their roots in the USA. The UK services were initiated in 1993, 1994, and 1991, respectively, and these companies are therefore classified as late entrants.

COLT was founded in 1992 by Fidelity Investments, following its experience in co-founding a local carrier in Massachusetts that provides services to business customers on its completely fiber optic network. This carrier offers both wire and switched services connecting users to long-distance carriers or to other user locations.

COLT was formed to provide similar voice and data transmission services to business users in London. Having been awarded a PTO licence in 1993, COLT constructed an initial 15-kilometer fiber optic backbone in the heart of the City of London before launching services on its network. Since its launch, COLT has extended its network east into the Docklands, south and west into Westminster, and into the West End, such that the network of fiber optic cable serves over 400 commercial and governmental buildings. All outgoing calls travel locally over BT lines, but are then routed nationally and internationally by COLT.

In June 1995, COLT was awarded a licence to construct a similar network in Frankfurt, Germany, which was launched there on 6 March 1996. The same month, it was also awarded a national UK telecommunications licence.

MFS Communications operates through its subsidiaries in telecommunications services (MFS Telecom, MFS Intelnet, MFS Datanet, and MFS International) and network systems integration (MFS Network Technologies). The fiber optic network is the core of MFS's operations. This is designed to connect business and government customers in the USA and Europe. Over the network platform, MFS provides a range of services, including voice, data, video, and interactive multimedia services.

MFS International was formed in 1993. Many of MFS's large corporate and financial customers are global companies. MFS expanded internationally to meet these customers' needs in international financial centers, as well as to serve the needs of European-based international companies.

In 1994, MFS International inaugurated services in London, Frankfurt, and Paris. The same year, MFS announced an agreement with British Telecom to interconnect networks in London. The London subsidiary also completed a fibre optic network that covered the London financial district and other parts of the city, including Docklands and Westminster. Furthermore, MFS obtained an international carrier licence, enabling the company to start work on a Pan-European fibre optic network that extends from the eastern seaboard of the USA into northern Europe, through Denmark, and directly into the UK and France.

By the end of 1994, MFS had received the authority to build a network in Frankfurt, Germany, the second largest financial center in Europe. MFS International was then granted a licence to become both a domestic and an international operator in Stockholm, and was authorized to build a network there. The same then happened in Paris. Further expansion in Europe will focus largely on the UK, Germany and France.

WorldCom has been in the telecommunications business since 1880, when the parent company of that time, ITT, was involved in laying out submarine cables around the world. In 1995, WorldCom was the fourth largest long-distance and international telephone operator in the USA and the third largest in international telephony in the UK (after BT and Mercury).

In the UK, WorldCom offers services that are also offered in the USA to all major international industry sectors and company sizes. The services combine international and national calls, as well as additional services, such as voice mail and fax facilities.

Scottish Telecom and Telstra Corporation are based in Scotland and Australia, respectively, and received public operator's licences in the UK relatively late. Both have long traditions in telecommunications, although they have somewhat different backgrounds.

Scottish Telecom has evolved from the telecommunications sector of Scottish Power, which, for over 50 years, has developed its own telecommunications network to securely support its electricity business.

With the financial backing of its parent company, Scottish Telecom has invested in a fibre optic network spanning Scotland's central belt. A network of fiber optic cables is in place, carried in part by Scottish Power pylons. This system has been augmented by forming strategic relationships with other providers, including the National Grid Company's telecoms arm, Energis, and COLT. This allows Scottish Telecom to offer its customers seamless services to the rest of the UK.

Scottish Telecom aims to build a customer connection network in Scotland that allows it to supply both large businesses and other customers. The network will be based primarily on fiber optic cables. Scottish Telecom is selecting partners for agreements for providing international connections.

The Telstra Corporation was established almost a century ago and is one of the leading Australian telecommunications companies. The corporation has primarily been present in the Asia-Pacific region, and it has offices in more than 30 countries.

With satellite technologies, over one million kilometers of fiber optic cable, and more than 5,000 telephone exchanges, Telstra links 95 percent of Australia's population to each other and to the rest of the world. In the increasingly deregulated telecommunications industry, Telstra is moving into markets such as the UK, Europe, and the USA, while further penetrating the growth markets of the Asia-Pacific region.

In the UK, it specializes in penetrating companies with significant communications traffic to Asia and the Pacific. Telstra offers understanding of the region and market knowledge, as it builds telecommunications infrastructure in countries like China, India, Indonesia and Vietnam. Initially, the focus is on services for telecom traffic of UK companies to and within the Asia-Pacific region.

Five late entrants into the UK operate telecom businesses with less relatedness to the original corporate core businesses. These operators are Atlantic Telecommunications, Orange Personal Communications Services, Racal Network Services, Scottish Hydro-Electric, and Videotron.

Atlantic was established in 1991 to develop telecommunications services, including voice telephony, using fixed radio-access radio-based technology. This technology will allow the deployment of a state-of-the-art digital communications network using radio frequencies in the local loop, without the need for underground construction, which can therefore be built at a fraction of the cost for hard-wired systems. The local loop is the part of the network that connects the subscriber to the local telephone exchange, historically the most costly part of any telephony network.

Atlantic was awarded a public operator's licence for the Strathclyde Region of Scotland in 1995. In March 1995, Atlantic was acquired by Caledonian Media Communications, a Scottish cable communications group with around 10 million pounds in turnover. Funds will need to be raised in order to develop the business beyond the concept stage.

> The acquisition of Atlantic allows us to test the viability, both technically and commercially, of our stated strategy of wishing to enter the telecommunications market. The concept will be fully explored in a controlled way in the Strathclyde Region and, if successful, will be developed in other parts of the UK, concentrating particularly on the areas in which we currently operate. We believe that deployment of a radio-based service will be more cost effective than a conventional telephone upgrade to an existing plant.
>
> We shall pursue the opportunities opened up by Atlantic in the field of radio-based telephony and multimedia services where the network can be built as consumer demand increases. We expect the most suitable markets for this service will be small to medium businesses and heavy usage domestic consumers, including the growing number of self-employed people working from home.

> We envisage that such facilities as home banking, colour fax, and database services will be in many more businesses and residential households within the next few years and it is our intention to provide these services. (Chairman Graham J. Duncan, Caledonian Media Communications, Annual Report, 1995)

Videotron, owned by North American companies, is another cable communications group that is entering the telecommunications market in the UK. The company was established in 1989 and holds licences to provide cable TV and telecommunications services to homes and businesses in London and Hampshire. (Later, Videotron became a part of Cable&Wireless Communications.)

Videotron uses a fiber optic network and serves more homes and businesses than any other operator in London (Bell Cablemedia, Cable London, Nynex, TeleWest and The Cable Corporation are other operators). The number of residential telephone customers grew by 42 percent in 1995, and the business customer base grew by 90 percent. Cable TV customers grew by 18 percent. At the end of the year, Videotron had 137,000 residential customers, of whom 103,000 took cable TV and 87,000 took telephone services, as well as 3,700 business customers.

In London, Videotron's strategy of acquiring contiguous franchise areas has facilitated construction of a single integrated network. This leads to economies of scale and allows Videotron to offer businesses the opportunity to network their premises.

The company owns and operates five Nokia exchanges to switch calls, and it connects to other operators, such as Mercury and BT. Videotron has links to five BT exchanges and to three Mercury exchanges and has established links to other cable operators.

Orange Personal Communications belongs to the Telecommunications business area of Hutchison Whampoa, a Hong Kong-based conglomerate with other activities in, for example, building construction, container traffic, retail, and energy. In the UK, the Orange network has made inroads in the digital cellular telephone market, and had a customer base of approximately 379,000 subscribers at the end of 1995. This corresponds to an approximate 26-percent share of the UK digital cellular telephone market and a 7-percent share of the overall cellular telephone market (Annual Report, 1995).

Racal Network Services belongs to the Voice and Data Communications sector of Racal Electronics. Other sectors are Defence Electronics, and Maritime and Industrial Services. Communications networks use a variety of digital and analog transmission media for local, national, and international communications.

From its inception as a company to run the governmental data network in 1989, its telecom operations have expanded and some years later had contracts with more than 300,000 customers in around 40 countries, but the government was still the most important customer of Network Services.

In 1995, Racal received a public telecommunications licence, and this complements the services it offers, enabling customer organizations to concentrate on their key business issues. Following the award of a licence, Racal can negotiate directly with multiple carriers on the customers' behalf.

Scottish Hydro-Electric has lately joined the telecommunications industry; the company holds a public telecommunications operator's licence in the UK. Scottish Hydro-Electric is an energy group that serves customers throughout the UK. Its roots are in the north of Scotland, and serving the community there is its first priority, although its energy business will be expanded primarily in England.

Entry strategy

In the UK telecommunications market, there are examples of entry modes based on organic developments, comprising both sole ventures and acquisitions, and strategic alliances formed for different reasons.

Kingston and Mercury are UK-based operators that were established relatively early following organic developments of sole ventures.

In the early 1890s the city corporation of Kingston-upon-Hull made plans for a municipal telephone system to compete with the established Post Office and the privately owned National Telephone Company's (NTC) networks. During the 1890s, the Post Office was threatening the NTC by taking over their national trunk network. The Postmaster General then set about establishing a local Post Office network to take on the NTC in rapidly expanding urban areas. To quicken the pace of competition, he also allowed local municipalities to borrow money on security of their general rates in order to set up their own telephone networks under a Post Office licence. There was one condition: the Postmaster General retained the right to buy back the municipal telephone systems.

To attempt to establish a council-owned telephone company was a decision few local authorities were prepared to make. Out of 1,334 local authorities, 55 expressed an interest. Some 13 applied for a licence, and six actually set up service (Brighton, Glasgow, Hull, Portsmouth, Swansea, and Tunbridge Wells). On 8 August 1902, Hull Corporation was granted its first licence.

In 1911, a decision was made that was to establish Hull's independence for years to come. The Postmaster General had taken steps to secure a UK monopoly of telephone services, buying out the NTC and many of the local authority-owned services which had fallen foul of poor planning or commercial failure, but he made Hull's bid for a new licence conditional on the purchase of the NTC network in the city. The council approved the purchase, and the sole municipally owned telephone company survived.

As a result of the Telecommunications Act of 1981, Mercury alone was granted a provisional operating licence, which was extended to a 25-year Public Telecommunication Operator's licence in 1984. This licence permits the operation of national fixed-link telephone lines.

When it entered into the market, Mercury chose a twofold strategy: First, it would offer the customer a combination of competitive prices and quality service and would be an innovative, proactive company. Second, it would enter a diverse range of markets. Although this policy spread resources over a wider area, it was deemed necessary in order to prevent BT from diverting its energy into Mercury's only chosen market and thus stifling growth.

However, deregulation of the market has let in many new competitors and has involved constant price undercutting. This means that it is no longer viable for Mercury to try to compete only on price. Another highly significant change is that customer expectations have risen greatly.

The changing conditions caused Mercury to alter its strategy. In the residential market, Mercury concentrates on customers in the target market whose bills are largely for long-distance and international calls. Regarding business customers, Mercury has moved from trying to diversify itself across the market to focusing on particular niches. Customization has become a high priority.

City of London Telecommunications (COLT) and MFS Communications exemplify the entry of foreign telecom corporations into the UK market, based on organic developments of sole ventures.

COLT was founded in 1992 by Fidelity Capital, a wholly owned subsidiary of Fidelity Investments, a mutual funds company, following its experience in co-founding Teleport Communications Boston (TCB). Founded in 1987, TCB is an alternative local carrier in Massachusetts that provides services to business customers through its completely fiber optic network, connecting users to long-distance carriers or to other user locations. COLT was formed to provide similar services to business users in London and was awarded an operators licence in April 1993.

MFS Communications Company was incorporated in 1987 in Delaware, USA, and prior to its public offering in May 1993, it was a wholly owned subsidiary of Kiewit Diversified Group. MFS operates through its subsidiaries in telecommunications, including MFS International, and network systems integration.

MFS International was formed in 1993 in order to meet the needs of global customer companies in international financial centers. In 1994, MFS International inaugurated services in London. By April, the London subsidiary MFS Communications Ltd had completed a fibre optic network that covered the London financial district and other parts of the city.

The path of Atlantic Telecommunications exemplifies an organic development comprising an acquisition. Atlantic was established in 1991 to develop telecommunications services, including voice telephony, using fixed-access radio-based technology. This technology relies on digital networks using radio frequencies, without the need for costly underground construction. Atlantic was awarded an operator's licence for the Strathclyde Region of Scotland in 1995.

In March the same year, Atlantic was acquired by Caledonian Media Communications plc, a Scottish cable television and multimedia communications group with around 10 million GBP in turnover and 130 employees for the fiscal year ending 31 March 1995. The acquisition allows the parent company to test the strategy of entering the telecommunications market.

The forming of strategic alliances is a mode of entry that frequently appears in the telecommunications markets. For instance, the Telstra Corporation of Australia often forms joint ventures with local governments and other corporations to develop and operate services and networks throughout the world. Telstra UK Ltd penetrates UK companies with significant communications traffic to Asia and the Pacific.

Mercury Personal Communications is a joint venture company formed in 1992 and equally owned by Cable & Wireless and US West, two international companies. The company is trading as One2One. This was the world's first operator of a digital personal communications network and was launched in September 1993. Its strategy involves building a brand, a company, and a network simultaneously. The objective was to reach positive cash flow in 1997/98, by which stage the network was to be national.

Moreover, strategic alliances are common means for operators to be able to offer their customers seamless services. The alliances of Scottish Telecom are one example of this. Scottish Telecom has invested in a multi-million-pound fibre optic network spanning Scotland's central belt. This system has been augmented by forming strategic relationships with other providers, including the National Grid Company's telecom arm, Energis, and COLT, thus allowing Scottish Telecom to offer its customers seamless services to the rest of the UK.

There are also other purposes for forming alliances, which the following example shows. In 1991, cable TV operators were licenced to offer not only television but also phone service across a local network. The Cable & Wireless response was to buy 20 percent of the holding company for Bell Canada Enterprises, which operates a UK cable television and telecom franchise. In return, Bell Canada bought 20 percent of Mercury shares. This strategic alliance offered mutual benefits: as mentioned, Mercury was able to bypass the use of BT lines for local calls, and Bell Canada was able to offer its customers cheaper long-distance and international calls. The alliance then developed into a complete merger and into the formation of Cable & Wireless Communications.

STRATEGIC STATES OF TELECOM OPERATORS IN THE UK

The first part of this section presents clusters of competing network operators in the UK market that initially followed certain combinations of business strategies in accordance with the strategic states model. Here, concentration, divergence, standardization, and adaptation constitute pure business strategies. Then, a discussion follows about applied competitive edges in terms of transmission technologies, pricing policies, and the use of distribution channels.

Strategy clusters

Figure 5.2 presents the strategic states of the network operators in the UK market in 1995. Customer types catered for are either companies or companies and private consumers. Regarding degree of customization, the operators were classified by me based on available information. In this dimension the extreme values are telephony, which is standardized in nature, and more or less unique communications services adapted to each customer. These solutions include telephony to some extent.

WorldCom is the single operator in the strategic state of concentration and standardization. Thanks to deregulation,

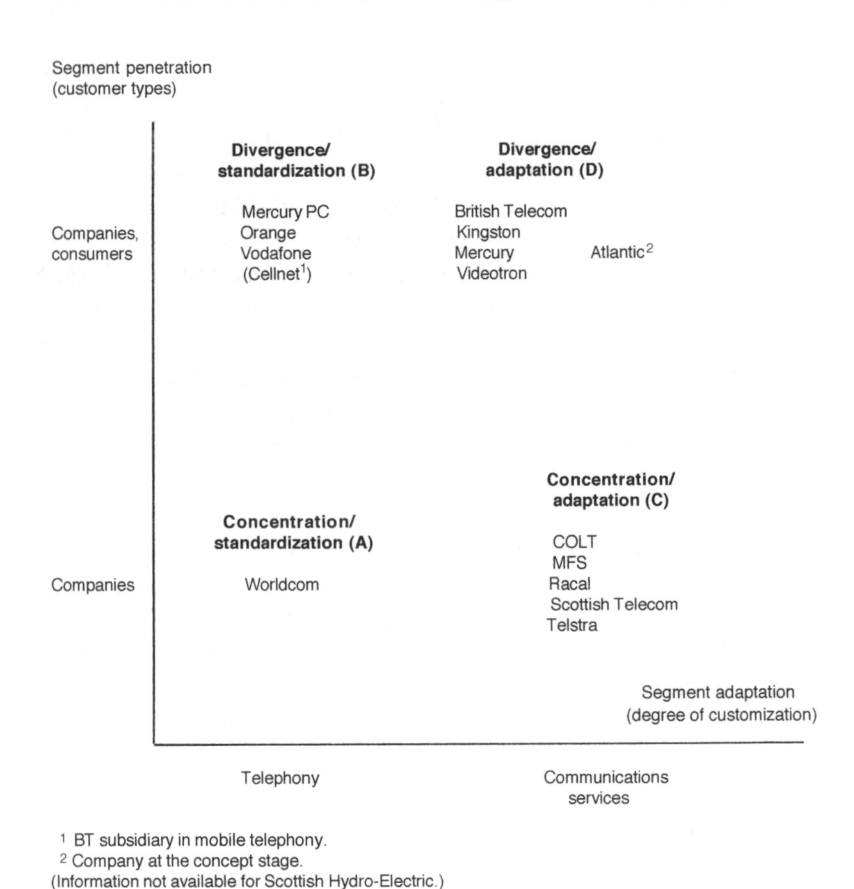

Figure 5.2 Strategic states of operators in the UK market in 1995.

WorldCom was able to start its UK operations in 1991. These are part of the original corporate core business, which can be traced to the activities of the previous parent company, ITT, in the 1880s. WorldCom is the fourth largest long-distance and international telephony operator in the USA and the third largest in international telephony in the UK (after BT and Mercury). The group's turnover was USD 2.2 billion in 1994, and the net loss was USD 122 million. In 1995, WorldCom employed around 7,000 staff members worldwide, with 100 people located in London.

In the first phase of establishment in the UK market, the company penetrated major corporations through services that had been available for similar organizations in the USA for some time. With the granting of a public telephony operator's licence, WorldCom then was able to offer these services to a wider range of companies. Although the strategy implies concentration on companies, it covers all major international industry sectors and company sizes. Customers are being directly connected to WorldCom's lines of fibre optic cables. The services are standardized in nature and deal primarily with international and national calls. Competitive pricing is the major marketing-based competitive edge put forward by WorldCom in order to become established in the strategic state to which I have assigned it.

The B group consists of mobile telephony operators with strategies that combine divergence and standardization. In principle, these companies penetrate any customer that needs a mobile telephone for communications. Pricing is an important marketing-based competitive edge, which can be exemplified by the charges of Mercury Personal Communications. This operator claims that it charges subscribers for local, national, and international calls at rates between 10 percent and 15 percent below what BT (and its subsidiary in mobile telephony, Cellnet) charges for the same calls.

Besides competitive pricing, the ability to treat distribution issues is a crucial condition for becoming established in this strategic state. The main reason is that standardized products imply that customers generally do not perceive any major differences between product features. Vodafone's elaboration of channels can be put forward as an example of the importance of distribution efficiency. Although the company expects an increase in retail distribution, it mixes service providers, specialist outlets, account managers and support from Vodafone's marketing department.

Operators in the C group are situated in a strategic state of concentration and adaptation. COLT, MFS Communications, Racal Network Services, Scottish Telecom, and Telstra penetrate companies by offering various customized communications services. All of these operators entered the UK market relatively late, and, except for Racal, telecommunications activities are part of the original corporate core businesses. However, Racal Network Services belongs to an electronics group.

COLT offers fibre optic private lines for all companies in London, and services comprise all levels of private line services, ranging from simple ring-down circuits to customized high-capacity links. Switched services provide access to both national and international operators. The company also acts as a broker and negotiates the most competitive prices for its customers for standardized telephony.

The fibre optic network is the technology-based competitive edge for MFS Communications. Its network connects business and governmental customers in the USA and Europe. MFS provides a range of services, including voice, data, video, and interactive multimedia services. MFS is based in the USA and has expanded internationally to meet the needs of large customers in international financial centers, as well as to serve European-based international companies. In 1994, MFS started its operations in London, Frankfurt, and Paris. The same year, it announced an agreement with BT to interconnect networks in London.

Racal Networks Services received a public telecommunications licence in 1995. This is a complement to the range of customized services they offer. Basically, Racal focuses on the idea of outsourcing. That is, Racal takes care of customer organizations' communications networks so that customers can concentrate on their key issues. Following the licence requirements, Racal negotiates directly with telephony operators on the behalf of customers. Since its inception as a company to operate the governmental data network in 1989, Racal has obtained contracts with many large companies.

Scottish Telecom has developed as a business related to the core of its parent company, Scottish Power. Although its telecommunications licence was granted in 1993, the parent has been developing its own telecommunications network for over 50 years in order to securely support its electricity activities.

The fibre optic network in Scotland is the technology-based competitive edge. This network has been complemented by agreements with other operators, including COLT, enabling them to offer seamless services to the rest of the UK and also internationally. Business customers in Scotland will be linked to the network by direct connections.

Even Telstra is a newcomer in the UK though with vast experience in telecommunications. The corporation was established a century ago and regards itself as one of the leading Australian telecommunications groups. Since World War II, Telstra has been providing services and networks to customers in the Asia-Pacific region.

As Telstra perceives deregulation is occurring, it enters key markets such as the UK, other European countries, and the USA, while continuing to penetrate the growth markets of the Asia-Pacific region. In the UK, Telstra initially penetrates companies that have significant communications with Asia and the Pacific region. The services include customized mobile telephony, voice mail, office exchanges, satellite digital networks, local training and expertise, and data networks. The company puts forward understanding and knowledge of the Asia-Pacific region as a main competitive edge in order to firmly establish itself in the strategic state of concentration and adaptation.

The established operators in the D group turn to both companies and private consumers who are interested in not only basic telephony but also in communications services involving solutions to specific demands. This means that BT, Kingston, Mercury, and Videotron are situated in a strategic state of divergence and adaptation, although it covers a certain degree of standardization. BT, Kingston, and Mercury entered the market early and are original telecommunications corporations with specific industry experience.

BT, the market leader, and Mercury penetrate companies and consumers in the national market with a mix of basic telephony and customized services. These companies also have international ambitions. In 1994/95, BT's turnover was 13,893 million pounds, and its profit was 1,736 million pounds. The corresponding figures for Mercury were 1,645 million pounds and 203 million pounds. For Mercury, the average five-year growth rates for turnover and operating profit were 23 percent and 15 percent, respectively.

In the residential market, Mercury concentrates on customers whose bills are largely for long-distance and international calls. Concerning business customers, Mercury has moved from trying to diversify itself across the market to focusing on particular segments. By the end of 1994, 99 of the top 100 British companies were customers, and there were around 1 million business exchange lines accessing the network. By March 1995, Mercury had 670,000 residential customers and 30,000 new customers a month.

Besides the competitive edge of building fibre optic cables, Mercury tries to negotiate marketing and operations agreements with cable TV operators in order to by-pass the use of BT lines for local calls.

In 1994/95, the turnover of the Kingston Group was 85 million pounds, with pre-tax profits of 7 million pounds. Kingston has replaced the diversification of earlier years with consolidation, which is its way of handling the technological and regulatory changes. The company remains municipally owned and has undertaken an expansion program based around its core telecommunications know-how.

The network of Kingston serves some 170,000 local customers, with access to BT and Mercury for all national and international calls:

> In many ways 1994 marked a turning point in our history when, for the first time, we were able to offer a range of local, regional, national and international services . . .
>
> In January 1995, Torch Telecom (a joint venture) began rolling out a broadband regional network targeting major corporate users in the Yorkshire region . . .
>
> Increasing competition in all our markets has sharpened our focus. Tightening cost control has been our policy. We must now consolidate our achievements and plan for sustained growth. (Chairman D. Woods, Kingston Communications (Hull), Annual Report, 1995)

The overall strategy is to grow businesses around the telecommunications core, while maintaining an investment program in its own public network. Kingston believes that the quality of its network management is its crucial competitive edge and underscores its importance in the delivery of customized services. Its aim is to grow its service portfolio to meet the demands of the business and local community. At the same time, Kingston expresses the need for standards across services.

Videotron is a late entrant, with its original core in cable TV operations. The company views itself as a local London and Hampshire business, serving more homes and companies than any other cable TV operator in these regions. In 1995, total revenues were 57 million pounds and net losses were 13 million pounds. The number of residential telephone customers grew by 42 percent in 1995, and the business customer base grew by 90 percent. Cable TV customers grew by 18 percent. At the end of the year, Videotron had 137,000 residential customers, of whom 103.000 took cable TV and 87,000 took telephone services, as well as 3,700 business customers.

Videotron sees its fibre optic network as a major competitive edge for the establishment. In London, acquiring contiguous franchise areas facilitates construction of an integrated network and exploitation of scale effects. Videotron connects calls to BT and Mercury exchanges and has also established links to other cable TV operators.

Low prices are regarded as a major marketing-based competitive edge, and Videotron estimates its price advantage over BT to be 10 percent for a typical private consumer or business customer in telecommunications. For the private customer, local voice calls are the main services, while business services include voice traffic and a mix of data and video services.

Competitive edge

City of London Telecommunications (COLT), MFS Communications, Scottish Telecom, and Videotron are operators that emphasize fibre optic transmission in stationary networks in order to become established in the UK market.

For example, before COLT launched its services, it constructed an initial 15-kilometer fiber optic backbone in the heart of the City of London area. COLT has since then extended the network into Docklands, south and west into Westminster, and into the West End, such that the network in the first quarter of 1996 included over 120 kilometers of fibre optic cable and served more than 400 commercial and governmental buildings.

Further, by April 1994, the London subsidiary of MFS had completed a fibre optic network that covered the London financial district and other parts of the city, including Docklands and Westminster. MFS also obtained an international carrier's licence, enabling the company to start work on a pan-European fibre optic network that extends from the eastern seaboard of the USA into northern Europe, through Denmark, and directly into the UK and France.

Besides Scottish Power and its Telecom subsidiary there are other utilities companies (for example, Energis and Norweb) that are trying to establish themselves in the telecommunications industry, by relying on their widespread existing infrastructures. These include internal telephony cables and a technique for wrapping fibre optic cables around existing earth wires on their electricity pylons.

Videotron is basically a cable TV operator that also provides telephony services. As we have seen, the company uses a fibre optic network and follows a strategy of acquiring contiguous franchise areas in order to construct a single integrated network in London. The aim is to reach economies of scale. Videotron might be viewed as a representative of cable companies that own their own local lines and are not tied to BT's access networks for residential calls.

Technological achievements allow fixed-link access service in the local loop to be deployed using fixed radio frequencies. The local loop is the part of the network that connects the subscriber to the local telephone exchange. By using radio frequencies, the need for costly underground construction diminishes.

Being one of the few companies building a network in the local loop gives Atlantic Telecommunications direct access to customers and the ability to exploit radio technology as a competitive edge.

Vodafone is another company that is applying radio technology in order to grow in the market. Although the UK is the home market for the Vodafone Group, its aim is to become a world leader in the field of mobile telecommunications products and services. The assortment comprises analog and digital radio network operation and service provision, value-added network services, radio paging, packet radio, and the design and manufacture of subscriber equipment and infrastructure.

Mercury Personal Communications was the world's first operator of a digital personal communications network, which was launched in September 1993. The network accesses every other telephone service, so customers can make and take calls to and from any number in the world. Mercury also expects to stimulate additional traffic by establishing "roaming" agreements in countries where system standards are compatible, enabling customers to make and receive calls abroad. The company wants to be a leader in the convergence of services, both between stationary and mobile services and between business and consumer users. The view is that as more people make use of mobile communications, business and personal usage is converging. The distinction is supposed to be between operators offering commodity air time and those trying to build more long-term relationships with customers.

In this study, pricing policy is seen as a marketing-based competitive edge. The following example deals with Mercury Communications' treatment of the pricing issue, and the example views pricing from a broader perspective.

When Mercury entered the market in 1984 and was allowed to compete with BT, it chose a twofold strategy (Annual Report, 1995). First, it would offer the customer a combination of competitive prices and quality service and would be an innovative, proactive company. This strategy was reflected in such practices as charging per second, as opposed to BT's practice of charging per unit. Mercury also offered fully itemized bills and breakdown of costs. Second, it would enter a diverse range of markets. Although this policy spread resources over a wider area, it was deemed necessary in order to prevent BT from diverting their energy into Mercury's only chosen market and thus stifling growth.

However, the market and the contenders within it have changed. Deregulation of the market has created opportunities for new competitors, frequently with price as their primary selling proposition. Hence, there have been price cuts in the market. (For example, the minutes used on the Mercury network grew by 19 percent in 1994/95, while revenue grew by only 12 percent.) This means that Mercury no longer views price as its main competitive advantage. Another significant change perceived by Mercury is that customer experience has increased greatly.

This changing situation caused an alteration in Mercury's strategy. In the residential market, the company concentrates on customers in the target market whose bills are largely for long-distance and international calls. Regarding business customers, Mercury has moved from trying to diversify itself across the market to focusing on particular niches.

National and international call charges continue to fall in real terms, and this forces operators to grow and/or reduce costs if the profit margins are to be kept constant. When it comes to pricing in this situation, reference is often given to the market leader, BT. For instance, Mercury Personal Communications charges subscribers for local, national, and international calls at rates between 10 percent and 15 percent below BT's rates for the same calls. Furthermore, Videotron claims that their price advantage over BT is estimated to be 10 percent for a typical customer. BT has responded with measures such as abolishing the highest rate for national long-distance calls and creating a simpler charging structure. The company has also reduced the costs of calls to the USA and Canada.

One reason for falling prices is the buying power of emerging brokers in the telephony market. COLT is one example; this company acts as a broker on behalf of its customers by using its bulk-buying power to negotiate the most competitive prices for its customers.

Efficient distribution structures might be a competitive edge as well. Vodafone is an illustrative case in this respect. The company markets its services through a variety of distribution channels. A network of service providers purchase air time wholesale from Vodafone and sell this and equipment directly to the end-user. A team of account managers supports the providers by offering training and information on the latest developments. As well as selling directly to the end-users, service providers can also market products and services through a network of specialist outlets and multiple retailers. The marketing department of Vodafone also provides support to this distribution channel.

Vodafone expects that the trend towards retail distribution will continue in the foreseeable future. In anticipation of this, 50 additional Vodafone Centers were opened during 1995, making a total of 115, most of which are located in major conurbations. Service providers added a further 260 branded retailers' outlets to this total and, in addition to these specialists, six high street multiples also sell hand-portables and connect subscribers to Vodafone services.

Videotron is an example of a company relying mostly on direct sales as a way of distribution:

> Our strategy involves a number of initiatives. Firstly we are focusing on sales and marketing, and have increased the size of our direct sales force. We are developing greater use of alternative sales techniques such as telemarketing and the use of external sales agencies. We have arranged to expand our presence in retail distribution outlets. For the first time we will mailshot and contact the existing base of homes twice in the forthcoming year whilst working hard to raise awareness of our services...
>
> We are also in the process of taking a fresh look at the segmentation of the residential market. This will allow us to take a targeted approach to bring considerable opportunities for growth in the future...
>
> We expect to launch free Voicemail services to all our customers in the new calendar year and to roll out number portability to all new customers, thereby eliminating the last major obstacle to conversion to our service for many would-be customers. (CEO Luis Brunel, Videotron, Annual Report 1995)

Figure 5.3 summarizes the examples discussed above of competitive edges in relation to the strategic states model.

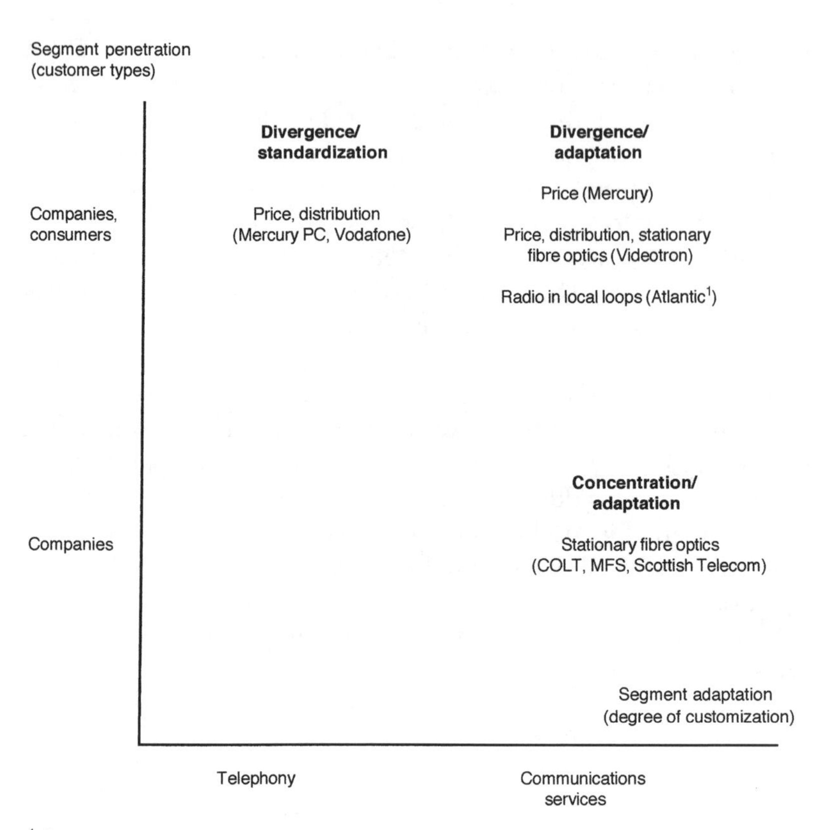

[1] Company at the concept stage.

Figure 5.3 Examples of competitive edges of operators in the UK market in 1995 as mentioned in the text.

SIX

Telecom Operators in Sweden

The measurements presented in Chapter 4 constitute the framework for the description of establishment processes and strategic states of network operators in the Swedish market presented in this chapter. Establishments are exemplified by company cases in the first section. The second section presents clusters of operators following certain strategy combinations, and competitive edges.

ESTABLISHMENT OF TELECOM OPERATORS IN SWEDEN

First, the establishment concept contains components valid for perceptions of two types of entry barriers: governmental regulations and the capital necessary for infrastructure investments in order to become firmly established in the market. The strategy competence component of establishment deals with relatedness between businesses and market experience. Finally, entry strategy concerns entry modes.

Perceptions of entry barriers

The precursor of Telia AB, Televerket, traditionally had a monopoly-like hold on the Swedish telecommunications market; however, during the 1980s, the market became more open for new entrants. In 1993, Telia AB was established as a parent company in the Telia Group. At the same time, a new telecommunications law was passed.

Telia made certain commitments in an agreement with the state that did not apply to other operators in the market. The agreement was in effect until the end of 1996 and included, among other things, constraints on the pricing behavior of Telia, social obligations without compensation, restrictions on interconnection fees, and permission for other parties to service switching systems earlier supplied by Televerket.

Telia's CEO perceives the deregulation process and affirms the changing character of the market:

> Since July 1, 1993, Sweden has been the world's most deregulated market with respect to telecommunications. In 1995, competition intensified strikingly, as numerous foreign operators entered the market. At present, in addition to Telia and many small niche operators, there are at least three companies with the ambition of covering all of Sweden. (CEO Lars Berg, Telia AB, Annual Report 1995)

One reason operators might have for trying to become established in the Swedish market is to get experience in breaking into former monopoly markets. NetCom Systems AB is the parent company of Tele2 and Comviq GSM, two entrants in Sweden, and its former Managing Director expresses this view:

> NetCom Systems AB is a young company in the telecommunications industry. At the same time we have much experience of breaching into monopoly markets. This has happened in Sweden which is one of the world's most deregulated markets...
>
> This means that NetCom Systems is well prepared to exploit the opportunities that deregulation and new technology create in the Nordic markets. (Managing Director Håkan Ledin, NetCom Systems AB, Annual Report, 1995)

The price constraints in Telia's contract implies difficulties for Tele2 in trying to expand its national telephony. Telia's position is regarded as one of the major challenges in the development of Tele2. The company's view is that competition in national telephony could be intensified if the Telia corporation was to be divided into one network company and one operational company that would compete with other operators in the market.

France Telecom initiated its operations in Sweden in 1992. Four years later the company took part in the creation of Global One, which is an alliance for international operations. Deutsche Telekom and US Sprint are also parties in the alliance.

> In Sweden we are able to study how a former monopoly reacts in the new situation. France Telecom can learn from Telia's mistakes, such as the defensive approach to moves of the competitors. (Director Monique Moulle-Zetterström, Svenska Dagbladet, 1996d)

In general, establishment in the market requires large infrastructure investments, (that is, investments in networks). This need for capital might be perceived differently in different situations. These are examples of the amount of capital needed for different types of investments:

- In 1995, Tele2 invested around 250 million SEK in its infrastructure for stationary telephony, and plans to invest the same amount annually in the coming years.
- Global One has invested approximately 1,000 million SEK in the Swedish market.
- Europolitan has invested more than 1,000 million SEK in its mobile telephony network in order to reach national coverage.
- In 1995, Comviq GSM invested 556 million SEK in its mobile telephony network (382 million SEK in 1994).

Frequently, capital is also necessary to cover operating losses. Europolitan, the mobile telephony operator, generated revenues of 752 million SEK in 1995, and its net loss was 400 million SEK. Operating costs were dominated by costs of interconnections to operators possessing fixed networks, sales commissions to resellers, and personnel costs.

MFS Communications reported significant losses in 1995 (although it is unclear if this also refers to the Swedish subsidiary); however, the company views its own network to be a key asset:

> During 1995, we've seen increasing evidence that the strategy we've been pursuing over the last several years is the right one. For the last seven years we've been building networks. It is now clear that those are the right assets, at the right place, at the right time. This last year, we've seen clear evidence of our ability to sell services over those networks. In particular, we've seen that MFS's early move toward combining local and long-distance — something we started three years ago — is the right strategy. (Chairman and CEO James Q. Crowe, MFS Communications Company, Annual Report, 1995)

Thus, network control is often viewed as a critical issue. For instance, Comviq GSM considers the control of its own network to be more favorable than renting capacity (Comviq GSM cooperates with Tele2 and with the Swedish State Rail Administration, and has access to their fixed fibre optic network.)

The building of networks itself does not generate revenues, but it does attract subscribers interested in the relevant price level. Tele8 intends to build its own network for the emerging DCS 1800 technology, and plans to offer cheap telephony services to private consumers. The company expects that more than half a million subscribers are necessary to break even.

Further, Telia is extending its area of investments and is broadening its focus to cover not only networks but also information services. In addition, a growing proportion of investments will be carried out abroad:

> The Group's new business objectives necessitate higher levels of investment and short-term costs in new ventures. Altough investment in Sweden will remain high in digitalization of the network, ISDN, ATM, the Internet, and mobile telephony, a growing proportion of investment will be made abroad. The new objectives and level of investment constitute a calculated increase in risk taken by Telia. To manage these risks, we are seeking partners and alliances among enterprises in the field of information media, we welcome financial partners in Telia Overseas, we are collaborating with our allies and AT&T in Unisource, and we are building our customer base before infrastructure in new markets. (CEO Lars Berg, Telia AB, Annual Report, 1995)

Strategy competence

The framework for this study suggests that strategy competence for a certain corporation is composed of its businesses' closeness to the original corporate core business and its experience in the market.

Figure 6.1 shows the classification of operators in Sweden regarding relatedness and market experience. Relatedness is covered by information on the closeness between the operation of telecom networks and the original corporate core business, while information on early and late entrance reflects experience in the Swedish market. Roughly speaking, operators established before 1991 are referred to as early entrants, while companies entering the market in 1991 or later are referred to as late entrants.

Telia is the largest provider of telecommunications services in Sweden and throughout the Nordic and Baltic regions. Residential, public, and business customers are penetrated by a wide range of

Relatedness	Market experience	
	Early entrants	Late entrants
Telecom operations are part of the original corporate core business	Telia Tele2 Comviq	Telenordia Global One MFS Telecom Finland Europolitan Tele8
Telecom operations are not part of the original corporate core business		Dotcom

Figure 6.1 Classification of operators in Sweden according to relatedness and market experience.

services. Seamless European and global services are being realized through the creation of alliances and partnerships.

The total Group revenues in 1995 reached 41,060 million SEK, with pre-tax profits of 3,227 million SEK. Restructuring of the organization is taking place in order to meet customer needs. In this process, five major market companies are being formed to cover the primary customer groups. These companies will be supported by units for the development and management of fixed and mobile telephony, information services, financial services, international operations and customer equipment.

Telia faces intensified competition and squeezed profit margins, which erode the profitability of the core telephony business:

> In 1995, the Group's total revenues grew 8 per cent. Traffic revenues in the fixed network climbed 2.2 per cent, mainly owing to traffic with the mobile network. At the same time, the volume of traffic rose 5 per cent, the difference being attributable to falling prices in 1995 . . .
>
> The market for mobile telephony continued to advance robustly, and Sweden reached a total penetration of 23 subscribers per 100 inhabitants. Serving more than 1.4 million mobile telephony customers, Telia leads the market, commanding a share exceeding 70 percent. (CEO Lars Berg, Telia AB, Annual Report, 1995)

Telia's reaction to this has been to broaden its markets geographically and to develop services that will generate revenues. In 1995, Telia, among other things, intensified its activities in neighboring regions in order to expand the home market, took part in the establishment of Unisource in the European market, and developed interactive services:

> To offset Telia's inevitable loss of market shares and profitability in its traditional domestic telephony operations, it was decided in 1995 to innovate and broaden the business focus. In addition to enhancing the degree of value added and continuing to refine the core business, the new business objectives entail investment in the field of information services, in other words becoming involved and earning money not only in the transport of signals in the network but also in creating, processing, storing, and presenting the information. Examples of these new fields are on-line shopping, interactive education, games and entertainment. (CEO Lars Berg, Telia AB, Annual Report, 1995)

Tele2 and Comviq GSM are the other two early entrants and belong to the original corporate core business of their corporation, the NetCom Systems Group. This Group deals with telecommunications and is engaged in, among other things, telephony and cable TV. In turn, this Group belongs to Kinnevik.

With around 20 percent of the Swedish market for international telephony and 6 percent of the market for long-distance traffic Tele2 is the largest private telephony operator in Sweden. The company also offers communications networks such as Internet, X.25, and Lan as well as leased lines. There are businesses for international and national telephony, communications networks, and the Internet. The major customer groups consist of companies and residentials for standard telephony and companies looking for customized communications solutions.

Comviq GSM is one of three competitors in Sweden in mobile telephony using GSM technology (the others are Telia Mobitel and Europolitan). On 31 December 1995, Comviq was serving 422,000 subscribers, which corresponds to roughly 40 percent of the Swedish GSM market. The turnover was 1,088 million SEK (449 million SEK in 1994), and the net loss after depreciation was 63 million SEK (140 million SEK).

Comviq GSM provides ordinary mobile telephony and additional services such as automatic answering and text messages.

The network is also capable of transmitting fax and data signals. Comviq has signed roaming agreements with operators in more than 30 countries. This means that customers are able to use their Comviq subscriptions in all of these countries.

The late entrants with telecom operations that belong to the core businesses of the corporations in question can be divided into three subgroups. First, there are two alliances between operators that strive for international leadership (Telenordia and Global One). The second subgroup is composed of companies that have their roots in other countries (MFS and Telecom Finland). Finally, the third subgroup consists of local companies (Europolitan and Tele8).

Telenordia is owned by British Telecom, TeleDenmark, and Telenor of Norway, all pure telecom corporations. Telenordia was established in 1995 and started to operate in the Swedish market. Its major goal is to be perceived as the best telecommunications operator in Sweden. Telenordia focuses primarily on companies' needs, and during 1995, many companies listed on the stock exchange became customers.

The company is building a nationwide digital network for telephony and data communications. The assortment is divided into national and international products and into services for telephony and data communications. As regards international services, Telenordia has access to the networks of the owners.

France Telecom Network Services started its operations in Sweden in 1992. Four years later this company became a part of Global One, which is an alliance between Deutsche Telekom, France Telecom, and US Sprint, a large American telecom operator. The worldwide network relies primarily on fibre optic and satellite technology. It covers around 60 countries, and exchanges are located in major cities in these countries.

The initial organizational structure was such that three managing directors, representing each of the parent companies, were to take joint decisions.

There were three parallel operations going on, without integration. I have now been appointed as the single Managing Director, and as a Swede I'm neutral. It would have been impossible with a German, Frenchman or American.

> It will be tough to reach the owners' objective of break even in 1998, but it is possible. The dominating costs stem from the buying of network capacity. In order to get cost effectiveness we need our own networks. (Managing Director Viesturs Vucins, Svenska Dagbladet, 1996c)

In the autumn of 1996, Global One had offices in the Swedish market in Stockholm, Gothenburg, Malmö, Ronneby, and Sundsvall, with a total of around 150 technicians. The Swedish network had 40 stations for connections. The company offers standardized and customized telephony and communications services. A co-operation agreement with Europolitan enables them to offer mobile telephony. The primary target groups consist of companies and authorities.

Metropolitan Fibre Systems (MFS) Communications is based in the USA and operates through its subsidiaries in telecommunications services and network systems integration. MFS International provides telecommunications services to business and governmental users in several major European metropolitan areas, as well as providing outbound international service from the USA and Europe. The company offers services over its own fibre optic network or through resale of international services in London, Frankfurt, Paris, Stockholm, and Zürich, and has announced plans to provide services in Hong Kong. In 1995, MFS began providing services in Stockholm, and the customer base includes the Stockholm options market. At the end of 1996, the MFS network covered almost the entire inner city of Stockholm.

The Telecom Finland Group (later Sonera) consists of four business sectors: Mobile Communications, Basic Networks, Value-Added Services, and Access Networks and Special Business Areas. In recent years Telecom Finland has extended its activities into international markets. The main target markets are regions adjacent to Finland, such as northwest Russia, the Baltic countries and Sweden. The company also operates in Belgium, the Netherlands, Hong Kong, and Germany.

Europolitan and Tele8 are local companies with national ambitions. Europolitan is a subsidiary of Nordic Tel Holdings AB, which was established in 1990 in order to operate mobile telephony networks in accordance with the GSM standard. The company was awarded an operator's licence in 1991, and its network was inaugurated in September the year after. In 1995, the turnover was 752 million SEK (278 million SEK in 1994) and the loss was 400 million SEK (a profit of 155 million SEK in 1994).

All activities concern the offering of mobile products and services following the GSM standard (although the company has received a licence to build a DCS 1800 network in Sweden, using higher frequencies for signals between telephones and base stations). At the end of June 1996, Europolitan served 225,000 customers, which was an increase of 52 percent since the end of 1995. During 1995, 78,000 new customers turned to Europolitan. This means that the company accounted for 13 percent of all new GSM customers in Sweden that year. Most Europolitan customers use their phones frequently. This implies large traffic volumes and corresponding demands on service resources. Variable costs per month for an average subscriber decreased by 40 percent between January and June 1996, compared to the first half of 1995.

Tele8 has formulated expansion plans. In March 1996, the company was awarded a licence for the next generation of mobile telephony technology: DCS 1800. (In August 1996, Comviq GSM, Europolitan, and Telia Mobitel also possessed licences, and these four operators have permission to manage nationwide networks.) The DCS 1800 technology requires more base stations than GSM technology does, but the networks have the potential for handling more subscribers. The Tele8 licence prescribes that all Swedish cities with more than 50,000 inhabitants are to be covered by the end of 1999 and that at least 50 percent of the total population is to be covered.

When we got the licence, several banks, financial institutions and venture capitalists expressed their willingness to join us. But our problem so far has been to choose financial partners. (Managing Director Carina Jonasson, Nordiska Tele8, Svenska Dagbladet, 1996b)

Enator Dotcom is the only operator in the Swedish market with telecom operations that are not part of the original corporate core business. Although communications services are the area of focus, sales and operations of telephony connections and leased lines are minor activities. Rather, data networks dominate the portfolio and the company regards itself as the market leader in systems for local networks. Office exchanges are also a major part of the portfolio, ranging from single exchanges to complete communications systems and including networks for buildings, radio and telephony connections, support systems, and so on.

The customer base consists of approximately 7,000 companies and organizations.

Middle-sized companies, large companies, and public administrations are the main target groups. Enator Dotcom tries to establish long-term relationships and considers repeat purchases to be essential.

Entry strategy

In the Swedish telecommunications market, there are examples of establishments based on organic developments of sole ventures and acquisitions and on strategic alliances formed for different reasons.

Tele2 and Comviq GSM are Swedish operators that were established relatively early following organic developments of sole ventures, although Comviq was an initial acquisition. Both companies are wholly owned subsidiaries of Netcom Systems AB, which, in turn, is a subsidiary in the Kinnevik Group.

Tele2 has its roots in Comvik Skyport AB, which was established in 1986. The original services comprised data transmission via satellite and the offering of equipment for the transmission of TV programs. In 1989, a long-term contract was signed with the Swedish State Rail Administration in order to build a common fibre optic network that would use the established railway network.

In 1990, the company name was changed to Tele2, and Cable&Wireless of England bought 39.9 percent of the company. Tele2 started to offer data network services and addressed these to companies, after which it also offered leased lines for national and international telephony. Broadscale marketing of public telephony services was initiated in March 1993, when companies and private consumers were offered cheap international calls by using a certain prefix. The corresponding service for national calls was introduced in October 1994.

Technological development and changing political attitudes towards deregulation at the end of the 1970s created opportunities for new establishments in the Swedish telecommunications market. Kinnevik started to build an analog NMT system for mobile telephony, which was operational in 1981 and managed by Comvik AB (this company was initially an acquisition); however, Comvik received permission from Televerket to engage only 20,000 subscribers. The marketing of the NMT system thus came to an end in 1996. In 1989, Kinnevik was awarded a national licence for the GSM system, and Comviq GSM AB started its activities in the autumn of 1992. Among other things, this company signed roaming agreements with operators in more than 30 countries. This means that customers are able to use their Comviq subscriptions in all of these countries.

Europolitan and Tele8 are also developing sole ventures organically, but these pure telecom companies were established at a later stage. Europolitan co-operates with Global One in the Swedish market.

> We have extended our network coverage, capacity and customer support function with a continuing speed. We have decided strategies for the development of technology and new services, and for the co-operation with other companies. Our underlying assumption is that customers will intensify their requirements on services. Access capability, no interruptions and several new services will be market standards. (CEO Tomas Isaksson, NordicTel Holdings AB, Annual Report, 1995)

In this study, two other companies represent foreign operators that were established in Sweden relatively late: Telecom Finland and MFS Communications. Telecom Finland to a high degree develops its international activities organically. The activities in Sweden started in 1994, where the company primarily addresses advanced network services for data communications, such as multimedia and Internet connections, and telephony to companies and organizations. In 1995, an operations and maintenance center, and a customer support function were established in Stockholm.

The company views the most promising area to be the provision of corporate network solutions for business customers (Annual Report, 1995).

In 1995, Telecom Finland focused on the development and expansion of Group companies and associated companies established earlier. One aim is to increase the volume of direct export of domestically developed services to international customers. This is motivated by the general growth of the telecommunications industry and by deregulation of European markets. Additionally,Telecom Finland derives income from international operators by providing its network for transit traffic:

> On the international side, in areas adjacent to Finland the focus was (in 1995) on the start-up and improved management of both fixed and mobile projects, while in more distant locations the activities centered around mobile projects. At the same time preparations were made for shifting our international emphasis from network operation and basic services towards service provision and value added network services. (CEO Aulis Salin, Telecom Finland, Annual Report, 1995)

The creation of Global One and Telenordia exemplifies that strategic alliances are a common mode of entering telecom markets, including the Swedish market. For instance, one aim of Telecom Finland is to enhance the value of the Group's international joint ventures by expanding both their networks and the business volumes carried by these networks. One example is a data network interconnection agreement that has been signed with MFS Communications. The interconnection of local networks is being extended from Finland to the rest of Europe and the USA. In the future, the two companies will offer network interconnection services. An objective of Telecom Finland is to raise the sales of the international joint ventures so that they account for one-third of the Group's net sales, while assuring good operating results as well.

STRATEGIC STATES OF TELECOM OPERATORS IN SWEDEN

The first part of this section presents clusters of competing network operators in the Swedish market. These follow specific combinations of business strategies (concentration, divergence, standardization, and adaptation) in accordance with the strategic states model. The second part presents applied competitive edges in terms of transmission technologies, pricing policies and the use of distribution channels.

Strategy clusters

Figure 6.2 presents the strategic states of the network operators in the Swedish market in 1995. Customer types are either companies or companies and private consumers. When it comes to degree of customization, I classified the operators based on available information. Here, the extreme values are standardized telephony and more adapted communications services. To varying degrees, these solutions contain telephony.

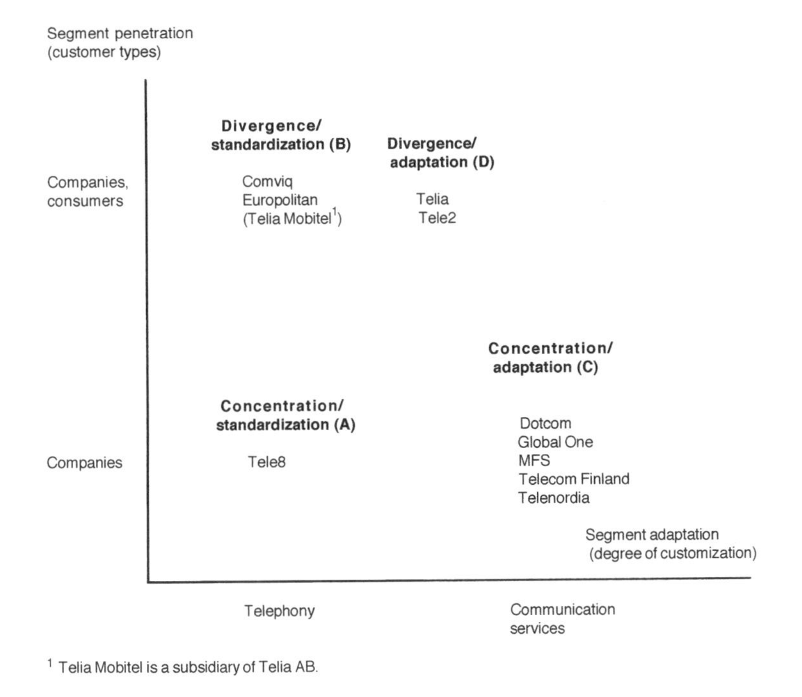

Figure 6.2 Strategic states of operators in the Swedish market in 1995.

The strategic state of concentration and standardization includes only one operator: Tele8. The Armstrong Group of the USA is the majority owner. This company group uses the public exchange of Tele8 as a point of connection to the rest of Europe. This means that over 90 percent of Tele8's traffic in 1995 was international traffic from Sweden over stationary lines. Tele8 concentrates particularly on small and medium-sized companies. The company realizes that price is the major competitive advantage in this market segment, but they are willing to differentiate the offer.

In March 1996, Tele8 was awarded a licence for the next generation of mobile telephony: DCS 1800, which uses higher frequencies for signals between telephones and base stations. This will probably lead to a divergence and a need to penetrate individual consumers.

Operators of mobile telephony using GSM technology and NMT are located in the strategic state that is characterized by a combination of divergence and standardization (the B group). The companies' operations are part of the original corporate core businesses, and both companies and consumers are being focused on. Competition is primarily based on price, and the operators compensate retailers for every new subscription.

Besides low prices, efficient distribution is crucial for the operators in the B group. This is mainly due to the varying character of their customers. Comviq GSM co-operates with chains of retailers and penetrates large customers directly. The network partly consists of the fibre optic network of the State Rail Administration, which means that dependence on Telia's network can be reduced. Furthermore, Comviq GSM and Tele2 belong to the same corporation and, thus, co-operate closely.

Europolitan was originally established in 1990 and is the youngest operator in the group. Links to the end-users exist through Europolitan Stores in a number of cities. This channel of distribution mainly addresses the large company market, while small companies and consumers are reached through co-operation with independent chains of retailers. Europolitan also co-operates with Global One in order to be able to offer complete telecommunications solutions.

Telia Mobitel is a subsidiary of Telia AB and belongs to the original corporate core business. Telia Mobitel develops and markets mobile communications services, and constructs and operates its own networks. At the end of 1995, Mobitel served almost 1,440,000 customers, representing around 70 percent of the total mobile telephony market (GSM and NMT subscribers). The corresponding figures for GSM subscribers were 460,000 and approximately 45 percent.

Operators in the C group entered the market relatively late and are located in a strategic state of concentration and adaptation. Only Dotcom is a local company, and the others represent altogether eight foreign operators. Dotcom is primarily a company of communications services, with only limited activities in telephony operations.

In the Swedish market, Global One (an alliance between Deutsche Telekom, France Telecom and US Sprint) offers standardized and customized telephony and communications services. An agreement with Europolitan allows them to the offer mobile telephony. The primary target groups are companies and public authorities.

Our goal in Sweden is to become the best alternative in competition with Telenordia and Tele2. The STATTEL contract (concerning deliveries to state authorities) will mean that we will attract even private companies. (Director Monique Moulle-Zetterström, France Telecom Nordic, Svenska Dagbladet, 1996d)

British Telecom, Tele Denmark, and Telenor of Norway are the owners of Telenordia. Telenordia concentrates primarily on companies' needs, and during 1995 many companies listed on the stock exchange became customers. Telenordia is building a nationwide digital network for telephony and data communications.

In Sweden, Telecom Finland mainly addresses advanced network services for data communications, such as multimedia and Internet connections, and for telephony to companies and organizations. The company views the most promising area to be that of providing corporate network solutions for business customers and has signed an agreement for data network interconnections with MFS of the USA. As is the case for the owners of Global One and Telenordia, in the home market, Telecom Finland possesses local access networks and penetrates private consumers, but for now has no intention of diversitying into this market segment in Sweden. Furthermore, The Telecom Finland Group also comprises operations of mobile communications.

MFS Communications provides company services in Stockholm, and its customer base includes the Stockholm options market. Low prices for local and international calls and price reductions for volume customers are major competitive measures in order to become established in the market. The technological reliability of the network is another advantage that MFS underscores in its marketing.

Operators in the D group turn to both companies and private consumers interested in not only telephony but also broader communications solutions. Hence, Telia and Tele2 are situated in a strategic state of divergence and adaptation, although the state covers a high degree of standardization. Both companies are early entrants, with their original corporate core businesses in telecommunications.

Telia sees residential, public, and business customers as equally important. These customer types are penetrated by a wide range of services. Seamless European and global services are being realized through the creation of alliances and partnerships; however, Telia faces a situation with intense competition and squeezed profit margins, so eroding the profitability of the core business. Telia's reaction is to broaden its markets geographically and to develop services that will generate revenues.

Tele2 belongs to the same company group as Comviq GSM does and has its roots in a company that was established in 1986. After 1989, Tele2 started to offer data network services to companies and later also provided leased lines for national and international telephony. Broadscale marketing of public telephony services was initiated in 1993 when companies and private consumers were offered cheap international calls. The corresponding service for national calls was introduced in 1994.

An interconnection contract with Telia allows Tele2 to penetrate residentials and small companies. These customer types are normally not profitable enough to be directly connected to the fibre optic network; however, Tele2 is exploring alternative access options in order to reduce its dependence on Telia's local networks. The use of the network of the State Rail Administration is complemented by applications of radio technology and fiber optics in local loops. These loops facilitate direct connections in geographical areas with many potential customer companies. In addition, Tele2 has signed a contract with Svenska Kraftnät (Swedish Power Networks) that allows them to install fiber optic cables in the existing power network.

Tele2 intends to keep its position as Sweden's largest private telecommunications operator, and a data communications actor. Its goals also include preserving its position as a leading Internet supplier.

Competitive edge

Tele2 uses the fibre optic network of the State Rail Administration to establish direct connections to large customer companies. Thus, this stationary network is a means of establishing itself in the Swedish market. Tele2 also applies radio technology and fiber optics in local loops.

Telenordia is building a nationwide digital network for telephony and data communications. Global One's worldwide network relies primarily on fibre optic and satellite technology. The Swedish network contained 40 stations in the autumn of 1996, but Global One is still dependent on Telia's networks:

> The dominating costs stem from the buying of network capacity. But I understand Telia's pricing policy. No one likes to lose market share. In order to get cost effectiveness we need our own networks. (Managing Director Viesturs Vucins, Global One, Svenska Dagbladet, 1996c)

MFS is expanding its fibre optic network, which is regarded as the core of its operations and which is designed to connect business and governmental customers in the USA and Europe. At the end of 1995, 5,720 buildings (2,754 in 1994) were connected either directly to the network or accessed through interconnections to other carriers' networks:

For the last seven years we've been building networks. It is now clear that those are the right assets, at the right place, at the right time. This last year we've seen clear evidence of our ability to sell services over those networks. (Chairman and CEO James Q. Crowcae, MFS Communications Company, Annual Report, 1995)

MFS and Telecom Finland have signed an agreement for the interconnection of data networks. Local networks in Finland might be extended to the rest of Europe and to the USA. In the future, the two companies will offer network interconnection services.

Costs of interconnections to operators possessing stationary networks are a heavy burden for mobile telephony operators as well. Europolitan primarily rents lines from Telia, while Comviq GSM tries to be independent of Telia, and instead uses the network of Banverket.

Competitive pricing is normally deemed crucial in order to become established in the telecommunications market, particularly for standardized telephony. Low prices for local and international calls and price reductions for volume customers are, for example, the major competitive measures that MFS underscores in its marketing. As mentioned, technological reliability of its network is another advantage that the company stresses.

Even Tele8's intention is to offer cheap telephony services, once its DCS 1800 network has been constructed. But low prices imply a need for high volume, and Tele8 expects that more than half a million subscribers will be necessary to break even.

In particular, we focus on small and medium-sized companies. It is difficult to attract large international companies in competition with operators like Telenordia and Global One. But, to sell telephony is like selling gasoline. You get the same product everywhere. But we try to differentiate and be more flexible than others. (Managing Director Carina Jonasson, Nordiska Tele8 AB, Svenska Dagbladet, 1996b)

Global One has the goal of becoming the best alternative in Sweden in competition with Telenordia and Tele2. In 1993, a general agreement was signed with the STATTEL delegation for delivery of data communications to state authorities. A general agreement for delivery of telephony services to state authorities was signed in 1996. (At the same time, STATTEL signed a similar agreement with Telia for the same type of services.)

The evaluation report says that Global One demonstrates the best technology and the best price. Our objective is to get 50 percent of the order. (Director Lars Persson, France Telecom in Sweden, Svenska Dagbladet, 1996a)

The general agreement includes an option to deliver to local communities. For instance, contracts have been signed with two Stockholm suburban areas and a major hospital. After this agreement, Global One expects that even private companies will be attracted.

Distribution is the second marketing-based competitive edge in this study. Tele2 treats the distribution issue in the following manner. The major customer groups consist of companies and residentials for standard telephony and of companies for communications solutions. Telephony services are offered by Tele2's own personnel for the major company accounts and through external distribution channels, such as telemarketing, for the small companies. Private consumers are being reached through direct advertising, TV commercials, and so on.

Comviq GSM, the mobile telephony operator, co-operates closely with leading chains of retailers. Each retailer receives a payment for every new subscription, and also a payment that is related to the customer's call volume. Comviq GSM supports the retailers through educational programs and sales support activities. TV commercials are used to a large extent to make the brand name well known. The company's own sales personnel penetrate the company market by offering specific company services. Low prices and high call quality in urban areas are being underscored as major competitive advantages for all customer types.

Europolitan AB and Europolitan Stores AB are subsidiaries of NordicTel Holdings. This means that the same company group controls both the construction and the distribution of its mobile telephony services. Seventeen Europolitan stores in eight cities distribute mobile telephony. Besides subscriptions for individual consumers, they offer subscriptions to large companies that allow these companies to integrate mobile telephony into their total communications system. In order to penetrate the mass market, consisting of consumers and small companies, Europolitan also co-operates with independent chains of retailers and carries out educational programs directed at retailers.

Enator Dotcom is a systems supplier and concentrates primarily on data networks and office exchanges. Middle-sized companies, large companies and public administrations are the main target groups. Enator Dotcom tries to establish long-term customer relationships, based on direct contacts with users, and considers

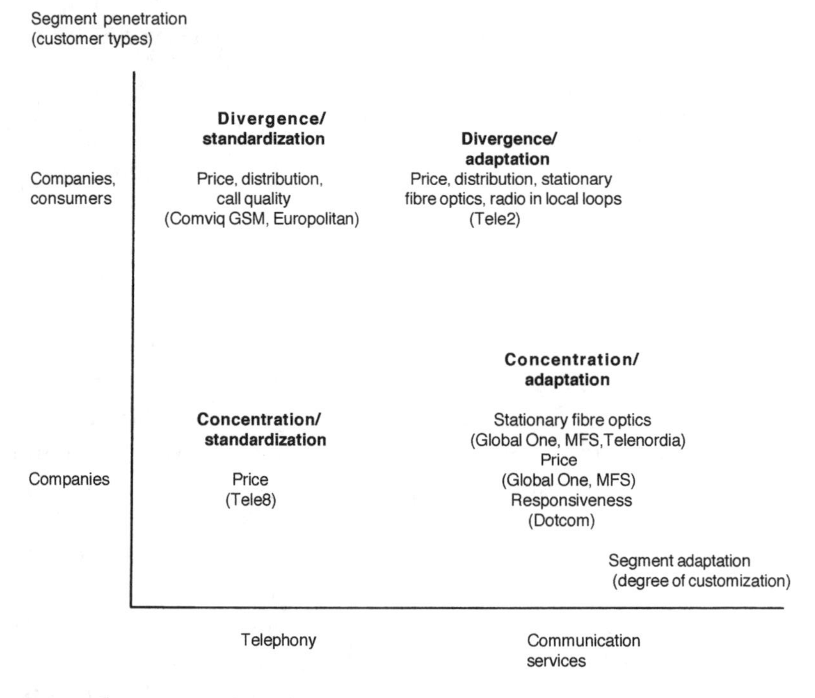

Figure 6.3 Examples of competitive edges of operators in the Swedish market in 1995 as discussed in the text.

repeat purchases to be essential. A delivery consists of these primary phases:

- Analysis of needs, which leads to general specifications.
- Projecting and systems design (equipment configurations, cable requirements, demands for financial resources, and time schedules).
- Installation (including tests of functions).
- Education and documents (customer-specific and general programs for education).
- Service (each customer chooses an appropriate service level).
- Financing (Enator Dotcom provides various financial solutions).

Figure 6.3 summarizes the examples discussed above of competitive edges in relation to the strategic states model.

PART III

Strategy Cases of Telecom Operators in Europe

INTRODUCTION

Following the strategy framework of this book, this part presents detailed case descriptions of strategies of representative UK and Swedish telecom operators. The descriptions comprise reconstructions of establishment processes and initial business strategies in the markets. Both written information from different sources, starting with information gathered in the cross-sectional study reported in Part Two, and information from personal interviews with leading managers were used as a basis for the descriptions. In order to secure external validity, respondents had the opportunity to comment on transcribed interviews and these were also checked against other information sources.

COLT Telecom represents niche operators and concentrates on high-volume telecommunications users, primarily in European financial and business centers, and offers highly adapted services to meet the customers' needs. This case highlights COLT's expansion into London, Frankfurt, Germany and other major European cities as a response to the liberalization of European telecommunications markets.

Cable&Wireless represents the large groups and strives to be a world leader in telecommunications. Here, the description initially treats the efforts to develop a corporate strategy based on the merger of four subsidiaries in the UK (Mercury Communications and the cable companies Nynex CableComms, Bell Cablemedia, and Videotron) and the creation of a common brand.

Mercury was already challenging BT, the UK dominant, in the beginning of the 1980s, and the establishment process and business strategy in Mercury's market before the merger is described. These aspects are also described for Videotron, which uses its cable TV network to offer telephony services in London.

When Tele2 of Sweden initiated the market establishment process as an early challenger to the state-owned operator, Telia, this dominant had complete control over international and national telephony in Sweden. One of the cases describes the experiences of Tele2 in challenging the market dominant. Tele2 belongs to the NetCom Systems group of Sweden. This relatively large group's efforts in developing a corporate strategy are also highlighted in a case description, covering topics such as intensified integration of group activities.

Comviq GSM, a mobile telephony operator, is another subsidiary in the group. This company's efforts to reach a strong position in the market are highlighted in one of the cases. A major reason for this choice is that the competition in the Swedish mobile telephony market is very intense. Mobile telephony has developed from a corporate market for car phones to a mass market for personal telephony. Norway, Finland, and Sweden are competing to be the country with the highest per-capita percentage of users in the world who are connected to the GSM system.

SEVEN

COLT Telecom: Expanding in Europe

COLT (City of London Telecommunications) aims to be the preferred supplier to high-volume telecommunications users and carriers in major metropolitan areas, primarily in financial and business centers. The prime target is large financial, media, corporate, and governmental users and carriers, for whom reliable telecommunications are crucial. In 1996, London and Frankfurt were the main metropolitan areas on which COLT focused.

> The global telecoms market is becoming functionally specialised. The long-distance carriers are adept at providing a competitive service across international barriers, but they all need robust local networks for access to customers. That is where COLT specialises. Beyond that there are telecom providers that will specialise in multimedia services or provide information services.
>
> As Europe opens up, COLT aims to build a local infrastructure in Europe's main business centers. COLT provides its customers with a wide array of access options for switching calls, nationally and internationally.
>
> It is important not to lose the focus as COLT expands. COLT therefore aims to concentrate on the business community only. It will also avoid becoming a vast bureaucracy, by allowing each metropolitan network to be controlled by a few key executives who have extensive local knowledge. (President Paul Chisholm, COLT Telecommunications, Newsletter, Issue 4, Autumn 1995)

COLT offers a wide range of local telecommunications services and competes with incumbent public telephony operators by emphasizing high quality and integrated services, primarily over fibre optic digital networks, to meet the voice, data, and video transmission needs of its customers. Switched services involve the transmission of voice, data or video to locations specified by end-users or carriers, while non-switched services include a fixed communications link between specified locations.

Essentially, all services, including direct Internet connections, that are based on COLT's network belong to the core business:

We stick to volume customers with our fibre optics network. The type of business customers depends on the city in question. In London, we focus on direct connections to financial businesses; these are generally early adopters of new technology. Smaller customers can use our COLT Connect device in order to be switched to other carriers. In London, we co-operate with around 30 other carriers...

Fibre optics and broad band technology are efficient in city networks, and our core technology stays. We don't perceive any major threat to fibre optics, and radio in local loops doesn't suit our ambitions. (Michael Bryan-Brown, Regulatory Affairs, COLT Telecommunications, 1 December, 1997)

The COLT Telecom Group plc consists of a holding company with subsidiary undertakings in London, Frankfurt, Paris, and Madrid (Annual Report, 1996). In 1996, the Group had telecommunications licences to provide services in Berlin, Hamburg, Munich, and Cologne, in Paris and in the Netherlands. The ambition is to create a uniform, pan-European identity under the COLT Telecom trade name, relying on common standards and network architecture.

1996 saw COLT implement its European strategy with the initiation of services on the COLT network in Frankfurt, and the award of additional licences in German cities and in Paris. The foundations are in place and COLT is well positioned to take advantage of the increasing liberalisation of telecommunications markets in Europe as the January 1998 date for full competition rapidly approaches.

The COLT networks facilitate the efficient flow of communications between customers and carriers, and serve to open markets previously controlled by monopoly suppliers. (Chairman Jim Hynes, COLT Telecom Group, Annual Report, 1996)

Tables 7.1 to 7.4 provide financial and operating information about the COLT Telecom Group plc.

Establishment in the European market

COLT was founded in 1992 by Fidelity Capital, a wholly owned subsidiary of Fidelity Investments, a mutual funds company, following its experience in co-founding Teleport Communications Boston (TCB). Founded in 1987, TCB is an alternative local carrier in Massachusetts that provides services to business customers on its completely fiber optic network. TCB provides both private wire and switched services connecting users to long-distance carriers or to other user locations.

Table 7.1 Profit and loss account of the COLT Telecom Group plc, GBP thousands (source: Annual Report, 1996).

	Year ended 31 December			
	1993	*1994*	*1995*	*1996*
Turnover				
– Switched	–	982	5.883	27,030
– Non-switched	32	1,483	3,274	7,791
– Other	–	–	33	158
Total turnover	32	2,465	9,190	34,979
Cost of sales				
– Interconnect and network	(71)	(1,097)	(5,162)	(24,396)
– Network depreciation	(63)	(783)	(2,039)	(5,044)
Gross profit (loss)	(102)	585	1,989	5,539
Operating expenses				
– Selling, administrative	(1,666)	(2,823)	(6,807)	(13,762)
– Other depreciation	(16)	(62)	(261)	(808)
Operating loss	(1,784)	(2,300)	(5,079)	(9,031)
Other income (expense)				
– interest receivable	–	78	118	531
– interest payable	–	–	–	(2,717)
– gain on (provision against) disposal of fixed asset investment	(2)	(501)	(295)	8
Loss for period	(1,786)	(2,723)	(5,256)	(11,209)

COLT was formed to provide similar telecommunications services to business users in London. Having been awarded a PTO licence by the Department of Trade and Industry (DTI) in April 1993, COLT constructed an initial 15-kilometer fiber optic backbone in the heart of the City of London before launching services on its network in October 1993. Commercial and governmental buildings were connected to the network. During the first years, COLT extended its network east into Docklands, south and west into Westminster, and into the West End.

Table 7.2 Balance sheet of the COLT Telecom Group plc, GBP thousands (source: Annual Report, 1996).

	Year ended 31 December			
	1993	*1994*	*1995*	*1996*
Fixed assets[1]	3,756	14,953	32,621	63,835
Current assets[2]	261	3,035	6,517	173,131
Total assets	4,017	17,988	39,138	236,966
Equity shareholders' funds[3]	(1,491)	9,231	6,468	80,263
Creditors	5,508	8,757	32,670	156,703
Total liabilities, capital and reserves	4,017	17,988	39,138	236,966

[1]Network infrastructure and equipment, computers, equipment, fixtures and fittings, and vehicles.
[2]Trade debtors, prepaid expenses and other debtors, cash and equivalents.
[3]Share capital, share premium, goodwill reserve, profit and loss account, contributed capital.

Table 7.3 Operating statistics of the COLT Telecom Group plc (source: Annual Report, 1996).

	1993	*1994*	*1995*	*1996*
Route kilometers (end of year)	24	72	118	198
– United Kingdom				129
– Germany				69
Buildings connected (end of year)	13	173	388	673
– United Kingdom				583
– Germany				90
Direct customers installed				580
– United Kingdom				537
– Germany				43
Indirect customers installed				2.319
– United Kingdom				
– Germany				0
Switched minutes (for year, millions)	–	14	115	514
– United Kingdom				514
– Germany				0

Table 7.4 Number of employees in the COLT Telecom Group plc (source: Annual Report, 1996).

	Year ended 31 December		
Employee categories	*1994*	*1995*	*1996*
Engineering and operations	25	50	103
Sales and marketing	9	24	58
Administration	8	20	40
Total	42	94	201

COLT intends to keep its focus on Europe, thanks to the size of European Union markets, their growth potential, and increasing liberalization of telecommunications.

According to the International Telecommunications Union, the five largest countries according to revenue are Germany, the UK, France, Italy, and Spain. These are all markets in which COLT intends to build networks.

Regarding potential cities to be penetrated, COLT continuously evaluates their attractiveness in terms of market potential and opportunities for becoming established in them. In practice, the process generally consists of these steps, once a specific city has been chosen:

1 Application for licence rights to build local infrastructure.
2 Discussions with, for example, local electricity and water authorities concerning appropriate city areas and streets to choose for the network.
3 Construction of restricted, often speculative, networks around the most interesting early adopters, such as leading banks.
4 Extensive penetration of potential customers.

In June 1995, COLT was awarded a licence to construct a fibre optic network in Frankfurt, which was launched there on 6 March 1996:

Chief among our milestones in 1996 was our success in Frankfurt. In just five months we promised to construct a network and have it ready for opening by the Mayor on 6 March. And that's exactly what we did. (CEO Paul Chisholm, COLT Telecom plc, Annual Report, 1996)

The concentration of large financial firms in Paris makes this city an attractive prospect for COLT. In July, 1996, COLT was granted a licence for the city and its nearby business district, La Defense. This licence allows COLT to provide switched and non-switched services within closed user groups.

In December 1996, COLT Telecommunications France SAS was one of the first in the country to be awarded a public network operator's licence. This licence is a complement to the closed-group licence and gives a number of benefits: the right to interconnect with France Telecom, access to public rights of way, access to telephone number allocations and rights to install and operate international facilities. As of January 1998, the licence will include full public voice telephony. COLT's initial services will cover an area including Paris and La Defense.

We have designed a network to serve Paris and La Defense, acquired space and have begun constructing an operations and administrative headquarters. We are in the process of finalizing the necessary approvals to begin network construction. (Directeur Generale Claude Olier, COLT Telecommunications France SAS, Annual Report, 1996)

COLT's future plans include establishment in Spain. A significant portion of that market is concentrated in two major cities, Madrid and Barcelona. COLT has designed a network for Madrid, has appointed a general manager, and is currently investigating options for obtaining licences and operating rights. Furthermore, entrance into the markets of Italy, Belgium, the Netherlands, and Switzerland are under consideration.

The process of acquiring licences is, however, not always straightforward:

There are countries where the introduction of a new telecommunications regime is proving a highly contentious issue. In such cases, COLT is actively lobbying for change on the principles of enhanced service and lower costs for the consumer. We may take more formal action if we feel that COLT is being deliberately obstructed by a country. For example, we have lodged a formal complaint with the European Commission because we believe that the Spanish Government is wrong to refuse us a licence to build and operate a network in Madrid. (Michael Bryan-Brown, Regulatory Affairs, COLT Telecommunications, Newsletter, Issue 8, Autumn 1997)

In order to link the city networks together, COLT has signed an agreement with Hermes Europe Railtel of Belgium. Hermes is building a pan-European inter-city fibre optic network. This is expected to make it possible for COLT to provide a fully integrated local and long-distance telecommunications service. The initial agreement involves the interconnection of COLT's local networks in London, Paris, and Frankfurt with Hermes's inter-city network (COLT Telecommunications Newsletter, Issue 8, Autumn 1997).

The network construction, though, puts a heavy strain on liquidity and capital reserves. The chief financial officer comments on this:

> The costs associated with the initial installation and expansion of COLT's networks in London and Frankfurt, including development, installation and early operating expenses, have been, and in new markets are expected to be, significant and will result in increasing negative cash flow. Negative cash flow is expected to continue in each of the company's markets until an adequate customer base and related revenue stream have been established. COLT believes its operating losses and negative cash flow will increase with the continued expansion of its networks...
>
> The company believes that its existing resources, together with internally generated funds and borrowings expected to be available under the bank facility will provide sufficient funds for the company to expand its business as planned. Under the company's current business plan, COLT expects to seek additional financing in 1999. (Chief Financial Officer Lawrence M. Ingeneri, COLT Telecom plc, Annual Report, 1996)

In December 1996, the parent company completed an initial public offering of 26,700,000 ordinary shares, listing on both the London Stock Exchange and the NASDAQ National Market in New York. Concurrent with the share offering, the company issued USD 314 million aggregate principal amount at maturity of senior discount notes. These were issued in the form of 314,000 units, each unit consisting of one 12 percent senior discount note and one warrant to purchase 7.8 ordinary shares.

On 9 January 1997, the underwriters of the company's initial public offering exercised their option to sell an additional 4,005,000 ordinary shares in the company. Net proceeds of the offerings totaled approximately GBP 180 million, which will be used to finance COLT's capital expenditure and working capital requirements in connection with the construction, expansion, and operation of its telecommunications networks in Europe, and to fund operating losses. Around GBP 14 million of the proceeds was used to repay short-term debts.

On 29 November 1996 the parent company entered into an agreement that established a credit facility of GBP 75 million with a consortium of banks. The bank facility comprises three branches and is, or will be, guaranteed by COLT's operating subsidiaries. The parent company has secured, or agreed to secure, the facility by a pledge of all of the stock of COLT's subsidiaries and a fixed and floating charge on substantially all of the assets of COLT and its subsidiaries.

This is a summary of major milestones of the market establishment process of COLT:

1992	COLT was founded by Fidelity Investment.
April 1993	COLT was granted a PTO licence allowing it to compete directly with BT and Mercury for voice and data transmission services within the Greater London area.
October 1993	COLT won its first customer.
February 1994	BT signed an agreement to allow COLT to ferry calls through its network. Since its inception, COLT has signed interconnect agreements with such other carriers as Scottish Telecom, Mercury, Telstra, US Sprint, AT&T, and Unisource.
September 1994	COLT teamed up with national long-distance operator Energis to offer joint services and to share some network infrastructure.
April 1995	COLT announced a new management structure to facilitate its expansion into Europe, based on strong local companies. It also plans to open an office in Frankfurt.
June 1995	COLT was granted a telecom licence to offer services in Frankfurt. DTI granted COLT a national UK telecom operator's licence. Under

	the terms of the licence, COLT would be able to expand its telecom network beyond London to other business centers such as Manchester and Birmingham.
August 1995	COLT completed the initial phase of its roll-out of a fibre optic cable network into the heart of London's West End. With it, the company plans to target leading media, advertising, and broadcasting companies.
January 1996	COLT launched COLT Connect, an indirect telecommunications service that will offer business users in Greater London savings of up to 30% on the cost of their long-distance and international calls.
September 1996	The parent company acquired the entire share capitals of its operating companies in London, Frankfurt, and Paris.
December 1996	The parent company completed an initial public offering of its ordinary shares on the London Stock Exchange and on the NASDAQ National Market in New York.

Initial strategy in the UK and German markets

COLT's market strategy in the UK has been to focus on the London market by continuously evaluating and adapting to the telecommunications needs of its target users and to offer comprehensive services by building and operating a high-availability network through the use of state-of-the-art technology. COLT aims to provide the same mixture of service, cost-effectiveness and flexibility that has been applied in London, to customers in Frankfurt, Germany, and to the rest of the UK as it starts to roll its services out to other UK and European cities.

COLT offers point-to-point fibre optic private lines for all businesses in London. Services include all levels of private line services ranging from simple ring-down circuits to high-capacity links. Switched services provide access to both national and international carriers. COLT also customizes configurations for businesses needing higher bandwidths or specialized arrangements.

With COLTConnect, all outgoing calls travel locally over BT lines but are then routed nationally and internationally by COLT. Acting as a broker on behalf of the user, COLT can use its bulk buying power to continually negotiate the most competitive prices for its customers.

On 31 December 1996, COLT's London network of 129,000 route meters provided switched and non-switched services to 537 directly connected customers in 583 buildings. The network embraces the major business concentrations of the City of London, the West End, the Docklands, and Westminster. For customers not directly connected to the network, COLTConnect is available.

> We pride ourselves on the quality of our service and our commitment to our customers. We demonstrate this commitment with a network availability in excess of 99.99% and a mean time to repair of only 56 minutes. (Managing Director Pat Hogan, COLT Telecommunications UK, Annual Report, 1996)

In March 1996, COLT initiated service in Frankfurt. This was five months after construction began on the network, and at year's end, this network covered 69,000 route meters, and connected 90 buildings. Customers include three major banks (Dresdner Bank, Commerzbank, and Deutsche Bank), financial services providers, including Bloomberg LLP, Knight Ridder, and Reuters, and carriers such as AT&T and Unisource.

In October 1996, COLT was granted new infrastructure licences in Frankfurt, Hamburg, Munich, and Berlin. Among other things, these licences allow interconnection with Deutsche Telekom at wholesale rather than retail rates.

COLT Telecom GmbH was one of the first applicants for licences under the new German telecommunications law, which became effective in August 1996. Later on, COLT was awarded licences for public switched voice services. These services came into effect in January 1998 and are among the first such licences to be awarded to a competitor of Deutsche Telekom.

As regards the co-ordination of COLT's activities in European markets, Mr. Bryan-Brown comments:

Today we are only 15 people here at the head office of COLT. Besides a common brand and the same services, we decentralize as much as possible to the local operating companies. Perhaps, when these grow, there will emerge a need for more central co-ordination. Essentially, the subsidiaries try to enhance the volume in the networks that we build. Pricing decisions are, of course, also local decisions. However, there's a price ceiling on all markets. Generally, this is decided by the local monopoly. But price is not our major competitive mean. Instead we put forward things like technological reliability and, for example, very quick repairs. (Michael Bryan-Bryan-Brown, Regulatory Affairs, COLT Telecommunications, 1 December 1997)

EIGHT

Cable & Wireless: Developing a Corporate Strategy

Cable & Wireless has its roots in the Eastern Telegraph Company, which was formed in 1872 under the guidance of John Pender, a Victorian entrepreneur. This company linked London to Hong Kong by laying out and managing communication through telegraph cables. In 1934, the company got its present name, Cable & Wireless. Together with the UK operation, the activities in Hong Kong still dominate the group.

The major ambition of the Cable & Wireless group is to be one of the world's leading international telecommunications groups. It provides, to both businesses and domestic users, services that include telephone, facsimile, telex, Internet, multimedia, and data transmission, making use of the most modern fixed-line and mobile technology available. In addition, the group provides and manages communication facilities and services for public and private customers and provides telecommunications consultancy worldwide. In 1997, the group had operations in over 50 countries and employed around 37,000 people. It possessed a fleet of 11 cableships and 18 subsea vehicles for the laying, burial, and maintenance of submarine cable systems.

The group primarily focuses on markets in Asia, Europe, the Caribbean, and the USA. In markets where it has fixed-line communications, its ambition is to offer mobile communications, and vice versa. Where the group offers data services, it wants to add video, multimedia, and cable TV. In fact, it considers itself to be a pioneer in offering not only domestic and international telephony, but also entertainment services over a single network.

Group turnover was 6,050 million GBP in the year ending 31 March 1997, while operating profit for the corresponding year was 1,538 million GBP (Annual Report, 1997). The parent company was introduced on the New York and London stock exchanges on 28 April 1997. Tables 8.1 and 8.2 provide further statistics.

The present CEO, Richard H. Brown, was appointed in 1996, and he expressed the belief that the highest priority should be to give value to investors. This implied a focus on the performance of the entire organization: definition of clear objectives, boldness, a bias for action, accountability, and efficient delivery systems.

Table 8.1 Turnover of the Group of Cable & Wireless plc by business, GBP million (source: Annual Report, 1997).

	Year ended 31 March	
	1997	*1996*
Group turnover:		
– International telephone services	2,804	2,560
– Domestic telephone services	1,553	1,442
– Other telecommunications services	1,165	986
– Equipment sales and rental	240	307
– Cableships and contracts	288	222
Total group turnover	6,050	5,517

Table 8.2 Turnover and operating profit of the Group of Cable & Wireless plc by geographical region, GBP million, year ended 31 March (source: Annual Report, 1997).

	Group turnover		*Operating profit*	
	1997	*1996*	*1997*	*1996*
Geographical region				
– Hong Kong	2,665	2,422	1,007	920
– Other Asia	118	106	10	11
– United Kingdom	1,718	1,698	317	183
– Other Europe	59	50	(38)	(35)
– Caribbean	603	548	194	179
– North America	621	477	41	41
– Rest of the world	327	273	7	12
– Interregional turnover	(61)	(57)		
Total	6,050	5,517	1,538	1,311

Performance targets were set, including double-digit revenue growth year after year and substantial annual improvements in productivity levels.

Cable & Wireless is a growth company in a growing market. The best measure of growth in a competitive marketplace is revenues and it will seek out creative and innovative ways to grow revenues. Management can also generate value by controlling costs and enhancing efficiency. Our intention is to widen the gap between costs and revenues year on year. One way to do that is by sharing learning experiences around the Group so that nothing is duplicated. (The Board of Directors, Cable & Wireless Communications plc, Annual Report, 1997)

Moreover, the CEO expressed a willingness to control the businesses, either by having controlling shares or by securing real operating influence:

This way we can ensure all investment goes into businesses we control and influence on behalf of our shareholders. Where this is not possible we will exit and reinvest elsewhere. (CEO Richard H. Brown, Cable & Wireless plc, Annual Report, 1997)

However, a dominating percentage of the growth previously came from businesses that did not carry the company name:

One reason we are regarded as the industry's best kept secret is that 80 per cent of our revenues have come from businesses which do not carry the Cable & Wireless name. To grow in a competitive world market place, we need to raise our profile; be one brand worldwide.

Our name is truly international with no geographic restriction. It also describes what we do. So instead of facing the market with different logos, we're moving towards a common identity, consistent with being a single enterprise that thinks and acts as one. (CEO Richard H. Brown, Cable & Wireless plc, Annual Report, 1997)

On 22 October 1996, Cable & Wireless announced that it had reached an agreement to create a large provider of integrated telecommunications, information and entertainment services by forming Cable & Wireless Communications plc (CWC) from a merger of the operations of Mercury Communications Ltd, Nynex CableComms Group plc, Nynex CableComms Group Inc., Bell Cablemedia plc and Videotron Holdings plc. The new UK company will offer local, national and international services in voice and data, together with multi-channel television, Internet services, and, in due course, interactive multimedia. "Cable & Wireless" is the brand name both for the group and for the UK operations. Graham Wallace was appointed chief executive officer of CWC.

CWC was divided into four business units to penetrate residentials, small businesses, corporations, and international markets (primarily Ireland), including wholesale activities: Consumer; Business; Corporate; International & Partners (wholesale).

A managing director was appointed for each business unit. These managers were to be responsible for revenues and costs and for accompanying issues, such as product decisions, pricing, and operational marketing. Further, the business units would be supported by functions of customer operations and network and staff in marketing, finance, communication, legal affairs, and so on.

However, the question of whether the central staffs or the business units were to be responsible for certain key activities became a major subject of discussion in the structural development. For example, the central marketing staff was initially responsible for common branding (that is, to inform the market about the existence of the CWC brand and related issues), while the business units took care of customer relations.

> Customer segmentation comes later. As regards the consumer market, for example, there seems to be significant differences between those customers demanding national and/or international telephony, those who demonstrate an explicit interest in high-tech based cable TV, and those consumers who are less willing to be engaged in high-tech cable TV. (Andrew Law, Research & Analysis, CWC, 27 March 1998)

The bringing together of four companies in the UK gives access to around six million homes and numerous businesses in a direct relationship not shared with any other provider. By running the combined traffic over one network, cutting out duplication, and standardizing services without losing sight of the need to tailor services to local markets, the intention is to raise efficiency. Furthermore, the merged company is expected to increase negotiating power with suppliers.

The previous Mercury subsidiary has a strong foothold in the UK corporate market, while the cable TV operations have strong positions in the residential market.

Around 90% of our revenues stem from telecom operations and the rest comes from cable-TV operations. I think that the residential market, mainly penetrated by the previous cable-TV companies, has the highest growth potential...

Together we will create the first integrated multiservice communications company in the UK. It is important to be in advance of BT. But how far on the value content side shall we go? For example, is it appropriate to produce entertainment? (Andrew Law, Research&-Analysis, CWC, 19 November, 1997)

The four brand names of the merged companies were simultaneously subsumed into the single "Cable & Wireless" name. As an integral part of launching the new brand name in the UK, a survey questionnaire was initially sent out to all customers in both the company and household markets. The response rate in the consumer market was 23 percent, while it was 5 percent in the company market. The information received primarily comprised relevant customer perceptions of services and related issues.

Before the merger, the Mercury brand name had a bad reputation in the company market. One reason was previous failures, such as the misleading efforts to launch payphones...

In connection with the merger we discussed two main alternatives: one brand name for the entire group, or separate names for well-defined markets, such as the corporate market and the residential market. The discussion led to the choice of "Cable & Wireless", a common brand name for the group...

Essentially, our profile is based on credibility. As regards services, we put forward flexibility and accessibility. Our intention is to be perceived as an inspiring company with personal relations to customers. (Andrew Law, Research&Analysis, CWC, 19 November, 1997)

Starting in June 1997, the first phase of launching the brand name in the UK included sending information about the merger directly to important corporate customers and opinion formers who were connected to the business press. Simultaneously, internal integration of processes, products, services, packages, and so on was started. This was combined with an extensive internal training program. After that, the process of bringing the Cable & Wireless brand name to the mass market, including both previous and potential customers, was initiated. The third and final phase consisted of a response to the survey by giving feedback to existing and potential customers who had participated in the survey. Thus, the campaign to launch the brand name consisted of these activities:

Phase 1 Brand launch

Timing:	1 June 1997 for 4 weeks.
Audience:	Customers and key opinion formers.
Objective:	To explain the merger and announce the new company/brand.
Proposition:	"Eyes, ears, mouth — they work better together."
Media:	Quality press and posters, limited direct mail.
Budget:	GBP 2.7 million.
Targets:	Spontaneous and prompted name recognition.

Phase 2 Brand launch

Timing:	15 September 1997 for 8 weeks.
Audience:	Existing and potential customers in consumer and business markets.
Objective:	To build awareness of the brand and demonstrate brand values.
Proposition:	The survey — "What can we do for you?"
Media:	TV, press, posters, direct mail.
Budget:	GBP 19.5 million.
Targets:	Spontaneous and prompted name recognition, recall of advertising, brand attributes.

Phase 3 Brand launch

Timing:	15 March 1998 for 3 weeks.
Audience:	Existing and potential customers in consumer and business markets.
Objective:	To respond to the survey and demonstrate brand values by introducing product solutions.
Proposition:	Feedback results — "Here's what we can do for you."
Media:	Direct mail, press, posters.
Budget:	GBP 4 million.

The total number of employees in the UK of the merged company in the first year decreased from around 12,500 to 11,500. Essentially, duplications were taken away in order to reduce costs (for instance, reduction of the number of call centers from 10 to three and a merger of five network control centers to two sites). Furthermore, one intention is to integrate the four different subscriber management systems of CWC into a single system.

Parallel to the initial rationalizations, CWC management identified a certain need for upgrading the long-distance fiber optic network, proceeding with the extension of the cable network, and the expanding of customer services in the consumer market. The latter expansion was for the purpose of significantly increasing revenues in the consumer market. The long-term costs associated with network investments and customer service expansion would, to a large extent, be financed internally, and, consequently, there was a need for higher profits. On the revenue side, increased volumes became the major objective, while further rationalizations would imply decreasing operating expenditures.

With the purpose of getting a picture of what was going on and of raising organizational efficiency, key activities and competencies within CWC were identified, starting at the end of 1997. Senior management then set priorities among key activities as a basis for further decisions. In this review of resources, each organizational unit calculated resource demands pertaining to different service levels, such as the current level, enhanced levels, and a minimum level.

MERCURY: CHALLENGING THE UK DOMINANT

Mercury Communications Ltd (later a part of Cable & Wireless Communications) had a number of subsidiary undertakings and was the parent company in the Mercury Communications Group. The Group's principal activity was the supply of telecommunications services and equipment. In fact, Mercury Communications was capable of offering a full range of telephone and data transmission services both nationally and worldwide.

Although turnover increased by 12.5 percent from 1994 to 1995, operating profits declined as a result of higher costs. The minutes used on the Mercury network continued to grow overall by 19 percent year after year. The average five-year growth rates for turnover and operating profits were 23 percent and 15 percent, respectively. Tables 8.3 and 8.4 provide further financial and operational information.

Thus, rationalization was initiated together with withdrawal from certain non-core activities, such as the operation of payphones. Despite the restructuring, volume continued to grow. For instance, at the end of March 1995, the number of installed lines stood at almost 2.5 million, up 48 percent from the previous year. The number of customer lines connected via cable operators reached over 715,000, an increase of 111 percent from March 1994.

Table 8.3 Profit and loss account of the Mercury Communications Group, GBP million (source: Report and Accounts, 1994/95).

	Year ended 31 March	
	1995	*1994*
Turnover		
– International telephone services	559.3	493.2
– Domestic telephone services	605.3	543.7
– Equipment rental and sales	146.2	154.8
– Other telecommunications services	344.0	279.3
Total turnover	1,654.8	1,471.0
Operating costs		
– Outpayments to other operators	634.0	525.2
– Employee costs	226.9	171.5
– Depreciation	240.4	209.2
– Other operating costs	420.1	330.4
Total operating costs	(1,521.4)	(1,236.3)
Operating profit	133.4	234.7
Profits on ordinary activities before taxation	73.2	261.6
Profit attributable to shareholders	43.9	218.8

The Group is now concentrating in markets where Mercury already has a significant share and will focus resources on establishing a strong differentiated position in customer sectors which have substantial growth prospects, and in which it is possible to provide a cost-effective and significantly improved level of quality and choice of services to customers. (Mercury Communications, Reports and Accounts, 1994/95)

In order to facilitate the expected growth, the Group reorganized itself around five customer-based business units aligned to key segments of the market and supported by customer service,

Table 8.4 Balance sheet of the Mercury Communications Group, GBP million (source: Report and Accounts 1994/95).

	Year ended 31 March	
	1995	*1994*
Fixed assets		
– Tangible assets[1]	1,974.0	1,859.3
– Associated undertakings, other investments	1.3	2.2
Total fixed assets	1,975.3	1,861.5
Current assets		
– Stocks	30.2	30.2
– Debtors	316.8	321.4
– Cash	23.4	130.5
Total current assets	370.4	482.1
Total assets	2,345.7	2,343.6
Creditors		
– Amounts falling due within one year	511.1	588.2
– Amounts falling due after more than one year	38.5	25.4
Provisions for liabilities and charges	64.9	1.1
Capital and reserves		
– Called-up share capital	850.4	850.4
– Share premium account	735.7	735.7
– Profit and loss account	142.5	141.6
Shareholders' funds	1,728.6	1,727.7
Minority interest	2.6	1.2

[1]Primarily cable, plant and equipment in network.

network, and product development. The intention was to tailor packages of added-value solutions specific to relevant segments. Another element in the strategy was to build upon the growth in customer lines cable operators. These were the units:

- International Business Services
  This unit served customers within the sectors of institutional finance, pharmaceuticals, oil and energy, manufacturing and logistics, travel, and transport.
- Corporate Business Services and General Business Services
  These units concentrated on serving large UK business customers.
- Business Enterprise Services
  This unit focused on providing packaged national and international communications services for small to medium-sized enterprises.
- Home Business Services
  This unit specialized in developing and packaging communications services for people working from home.
- Partner Services
  This unit worked closely with other telecom providers, such as cable companies and international carriers.

Establishment in the UK market

Mercury Communications was formed in 1981 as a partnership between BP, Cable & Wireless and Barclays Merchant Bank in order to compete with BT; however, by the end of 1984, Cable & Wireless had become the sole owner by acquiring all the shares.

As a result of the Telecommunications Act of 1981, Mercury alone was granted a provisional operating licence, which was extended to a 25-year Public Telecommunication Operator's licence in 1984. This licence permits the operation of national fixed-link telephone lines.

In 1991, cable TV operators were licensed to offer not only television but also phone service across a local network. The C&W response to this was to buy 20 percent of the holding company from Bell Canada Enterprises, who operate a UK cable television and telecom franchise. In return, Bell Canada bought 20 percent of Mercury shares. This strategic business alliance offered mutual benefits: Mercury was able to by-pass the use of BT lines for local calls, and Bell Canada was able to offer its customers cheaper long-distance and international calls.

However, Mercury was not at all satisfied with the competitive behavior of BT:

> One obstacle in our path continues to be over-complex and increasingly unworkable regulation. This now threatens to impede the growth of competition, with customers as the losers. We are fast approaching regulatory failure, which only fundamental reform can resolve. Our recommendations for this reform require increased powers for the Director General of Oftel to curb anti-competitive behaviour and the removal of artificial barriers to competition. (Chief Executive Peter Howell-Davies, Mercury Communications, Annual Report 1995)

These are some milestones in the market establishment of Mercury Communications:

1981 Mercury Communications Ltd formed as a consortium by Cable & Wireless, BP, and Barclays Merchant Bank.

1982 Mercury granted its first operating licence.

1983 First London communications links established. London–Birmingham microwave route completed.

1984 Current licence granted under Telecom Act 1984. First Transatlantic leased services launched.

1985 Business services to New York and Hong Kong launched. Cardiff connected to Mercury network. Oftel determines terms for transferring calls between Mercury and BT networks.

1986 Agreement with AT&T for direct public services to the USA. Start of national public services.

1987 Direct services agreement with Australia. First European agreement for direct international services with Italy. Hull Telephone Company agrees to connect with Mercury.

Mercury connects network to cable TV operator (Windsor).

1988 Public service to Japan begins. Agreement with Deutsche Bundespost for direct services between Mercury and the Federal Republic. Network reaches Aberdeen and Dundee.

1989 Network extended to southern coast of England. Agreement for direct services between Mercury and France Telecom following completion of three undersea cables between the UK and France.

1990 Launch of Mercury's managed data network service. Interconnect agreement links Mercury and BT public data services. Agreement sets terms for telecom alliance between Mercury and all US West-backed cable TV companies.

1991 Network reaches East Anglia. 100,000th residential customer. Mercury Personal Communications awarded PCN licence. Award of contracts to lay cables to N. Ireland.

1992 Mercury Paging merges with InterCity Paging and becomes number two in the market. Mercury launches first National Virtual Private Network in the UK.

1993 A transatlantic optical fiber optic cable reaches Wales. Network extended to cover 90 percent of the UK population. ISDN routes more than doubled, reaching 22 countries. Mercury signs marketing and operations agreement with cable TV operators. Mercury launches One2One, a mass-market mobile service. Mercury wins contract to run the governmental long-distance telephone network and the police national network. Four major services are launched.

1994 Nine major services are launched. Mercury Calling Card joins forces with American Express. Mercury announces 200 million-pound investment in intelligent networking, the first part of a worldwide, billion-pound initiative by parent Cable & Wireless. Second directory inquiry center announced in Glasgow. Mercury takes a stake in the World Health Network and wins UK's largest corporate paging contract to supply British Rail Telecom.

1995 Four major services are launched. Welsh Water appoints Mercury and Rothwell to provide communications for their call center. Mercury outsources directory inquiries to American operator, Excell. Mercury wins wide area network contract to improve communications between offices of the Department of Transport. Mercury sells 75% of its Customer Premises Equipment division to Siemens.

Initial strategy in the UK market

At the time of its entry into the market, Mercury chose a twofold strategy. First, it would offer the customer a combination of competitive prices and quality service and would be an innovative, proactive company. This strategy was reflected in such practices as charging per second, as opposed to BT's practice of charging per unit. Mercury also offered fully itemized bills and breakdown of costs.

The second element of the strategy was to enter a diverse range of markets. Although this policy spread resources over a wider area, it was deemed necessary in order to prevent BT from diverting its energy into Mercury's only chosen market and thus stifling growth.

However, the market and the contenders within it changed. Deregulation of the market let in a wave of new competitors whose primary selling proposition was their price. Consequently, the market saw constant price undercutting, while the basic telephony prices fell. This meant that it was no longer viable for Mercury to try to compete primarily on price. Another highly significant change was that customer expectations increased greatly.

Thus, an internal evaluation of market policies and the recognition of a changing market environment caused Mercury to alter its strategy. In the residential market, Mercury concentrated on customers in the target market whose bills were largely for long-distance and international calls. Regarding business customers, Mercury moved from trying to diversify itself across the market to focusing on particular niches. Mercury's customer service became based on responsiveness to customer needs, ease of use, flexibility and customization for each customer.

By the end of 1994, Mercury had 99 of the top 100 British companies as customers for its telephone and data transmission services nationally and worldwide. By March 1995, there were more than 1 million business exchange lines accessing the network. By the same date, Mercury had 670,000 residential customers and 30,000 new customers a month.

Mercury regarded its main competitors to be BT, Kingston Communications, and City of London Telecommunications (COLT); however, new competitors emerged in the form of cable companies, digital radio-based operators, and electricity companies.

Cable companies make a big impact on the residential market. They own their own local lines and do not have to use BT's network for those calls. Some cable companies linked into the Mercury network and then offered competitively priced long-distance and international calls. It has been estimated that by the year 2002, the cable companies could control at least 7 percent of the telephony market.

The Ionica company was granted its PTO licence in 1993. It is one of a number of operators that have based its network on digital radio, which offers a better quality of line than cellular networks offer.

The utility companies have become involved in the telecom industry because they already have a large established infrastructure. This comprises the telephony cabling used on their own private networks and a system of wrapping fibre optic cables around existing earth wires on their electricity pylons.

For example, Energis is the wholly owned subsidiary of the National Grid Company and came on-line in 1994. Energis's network consists of 350 kilometers of fiber optic cable, which runs alongside their 7000-kilometer electricity network. The focus of their marketing is in the business market, but Energis continues to run trials in the residential market.

Norweb was granted a licence in 1994 and operates in the northwest of England, but it also offers private leased lines in other parts of the UK. Similar to Energis, Norweb has chosen to concentrate on the business sector.

VIDEOTRON: OPERATING TELECOMMUNICATIONS AND CABLE TV IN LONDON

Videotron Holdings plc was, until the creation of Cable & Wireless Communications, the largest integrated cable television and telecommunications operator in London. The Group holds licences from the Department of Trade and Industry and from the Independent Television Commission to construct and operate broadband cable networks providing a range of cable television and telecommunications services within certain franchise and licence areas in London and Hampshire.

The parent company's principal subsidiaries, all of which were wholly owned, were Videotron Corporation Ltd, Videotron Hampshire Ltd, Videotron West London Ltd, and Videotron South London Ltd. Videotron Corporation was engaged in the provision of administrative services, while the other subsidiaries provided cable TV and telecommunications services.

The Group's licences included over 1.3 million potential homes and approximately 138,000 potential businesses. The Group offered cable television, residential telephone, and business telecommunications services in the areas of West and South London and in Hampshire on the southern coast of England. The operations included telecommunications services in the City of London and in the City of Westminster, the geographical, commercial, and financial center of Greater London.

On 31 August 1996, the Group had 134,975 households subscribing to its cable television service. There were 123,133 residential telephony customers and 18,654 business exchange lines connected. The networks had reached 58 percent of the total franchise area required by the Group's licence obligations.

Tables 8.5 to 8.8 provide further financial and operational statistics.

Table 8.5 Cumulative statistics of Videotron Holdings plc as of 31 August 1996 (source: Report and Financial Statements, 31 August 1996).

	South London	*West London*	*Hampshire*	*Total*
No. of premises passed	276,601	293,469	107,869	677,939
No. of residential telephone lines	47,872	39,609	35,652	123,133
No. of residential cable TV customers	55,661	59,005	20,309	134,975

Table 8.6 Operational statistics of Videotron Holdings plc (source: Report and Financial Statements, 31 August 1996).

	Year ended 31 August	
	1996	*1995*
Cable television		
– Average monthly revenue per customer, GBP	22.69	21.29
– No. of customers	134,975	103,361
– No. of customers gained in year	31,614	15,980
– No. of homes passed	651,168	561,556
– Penetration based upon homes passed	20.7%	18.4%
Residential telephone		
– Average monthly revenue per customer, GBP	31.19	31.18
– No. of customers	123,133	86,506
– No. of customers gained in year	36,627	25,729
– No. of homes passed	595,703	485,172
– Penetration based upon homes passed	20.7%	17.8%
Residential market penetration	27.7%	24.3%
Business telecommunications		
– Average monthly revenue per exchange line, GBP	54.06	53.58
– No. of exchange lines	18,654	12,448
– Average monthly revenue per customer, GBP	202.86	169.47
– No. of customers	4,624	3,664
– No. of exchange lines per customer	4.0	3.4
– Businesses passed	26,771	23,532
– Penetration based upon businesses passed	17.3%	15.6%

Table 8.7 Consolidated profit and loss account of Videotron Holdings plc, GBP thousands (source: Report and Financial Statements, 31 August 1996).

	Year ended 31 August	
	1996	*1995*
Turnover		
– Cable television	32,443	24,498
– Residential telecommunications	39,226	27,277
– Business telecommunications	10,088	5,715
Total turnover	81,757	57,490
Cost of sales	(33,874)	(27,506)
Gross profit	47,883	29,984
Administrative expenses	(30,882)	(23,041)
Non-cash technical assistance fees	(2,815)	(1,833)
Depreciation and amortization	(16,435)	(12,485)
Operating loss	(2,249)	(7,375)
Net interest expense	(14,181)	(5,833)
Tax credit on loss on ordinary activities	142	47
Loss on ordinary activities after taxation	(16,288)	(13,161)

Establishment in the market in London

British cable operators enjoy the freedom to offer both telephone and television services within licence areas. Close to 90 cable TV franchises offer their customers telephone services and more are entering the market each month (Annual Report 1995). Cable networks are expected to cover 75 percent of the country's population within a decade. Around 40,000 kilometers of fibre optic cable had been laid out by the end of 1995. The total civil engineering investment in the national network is being funded exclusively by the private sector. The cable networks sector has received vigorous backing from the global investment community.

Table 8.8 Consolidated balance sheet of Videotron Holdings plc, GBP thousands (source: Report and Financial Statements, 31 August 1996).

	Year ended 31 August	
	1996	*1995*
Fixed assets		
– Intangible assets[1]	6,849	7,370
– Tangible assets[2]	505,431	390,218
– Other investments	754	979
Total fixed assets	513,034	398,567
Current assets		
– Debtors due after more than one year	8,478	13,941
– Debtors due within one year	26,070	18,405
– Investments	16,240	9,278
– Cash and deposits	–	55,061
Total current assets	50,788	96,685
Total assets	563,822	495,252
Creditors, amounts falling due		
– within one year[3]	53,423	43,568
– after more than one year[4]	317,637	245,815
Capital and reserves		
– Called-up share capital	13,795	13,699
– Share premium account	239,827	236,742
– Goodwill reserve	(20,222)	(20,222)
– Profit and loss account	(40,638)	(24,350)
Total capital and reserves	192,762	205,869

[1]Franchise application costs and deferred development expenditure.
[2]Telecommunications network, subscriber electronics and equipment, vehicles, fixtures, fittings, plant machinery, land, buildings and leasehold inprovements.
[3]Property loan, bank overdrafts, finance lease obligations, trade creditors, amounts owed to principal shareholders, taxation and social security, other creditors, accruals and deferred income.
[4]Senior loan, property loan, finance lease obligations, senior discount notes.

Videotron viewed itself primarily as a local company. Videotron reached and served more homes and businesses than any other operator in London. Bell Cablemedia, Cable London, Nynex, TeleWest and The Cable Corporation are other operators.

In London, Videotron's strategy of acquiring contiguous franchise areas facilitated construction of a single integrated fibre optic network. This led to economies of scale and allowed Videotron to offer networking of business/premises. Before the end of the decade, Videotron planned network investments involving over 1 billion pounds, besides the 400 million pounds the company had already spent (Annual Report, 1995).

In 1995, Videotron owned and operated five Nokia exchanges to switch calls and to connect to other operators, such as Mercury and BT. Videotron had links into five BT exchanges and into three Mercury exchanges, and had established links to other cable operators.

Initial strategy in the market in London

The broadband cable networks of Videotron made possible the provision of a range of cable television and telecommunications services within the relevant franchise and licence areas in London and Hampshire. The customer groups consisted of residential telephone and cable television customers and of business telecommunications customers. The range of services was somewhat broader for the latter market segment and included voice traffic and a mix of data, voice, and video services.

Videotron's number of residential telephone customers grew by 42 percent in 1996 (42 also percent in 1995), and the number of cable TV customers grew by 31 percent (18 percent in 1995). The business customer base grew by 26 percent (90 percent in 1995).

Cable operators view their fibre optic networks as an advantage they have over longer established telecommunications operators because the networks they are constructing are state of the art. Instead of conventional copper cable, which links the local exchange to each customer's premises, Videotron's fibre optic cable offered the advantage of greater capacity, better signal clarity, and a higher transmission speed.

Videotron's price advantage over BT was estimated to be 10 percent for a typical customer, which included free local voice calls to other Videotron customers for residential customers. The same type of price advantage was available to business customers.

The chief executive officer described the salient features of Videotron's strategy in the market:

> Our strategy involves a number of initiatives. Firstly we are focusing on sales and marketing, and have increased the size of our direct sales force. We are developing greater use of alternative sales techniques such as telemarketing and the use of external sales agencies. We have arranged to expand our presence in retail distribution outlets. For the first time we will mailshot and contact the existing base of homes twice in the forthcoming year whilst working hard to raise awareness of our services...
>
> We are also in the process of taking a fresh look at the segmentation of the residential market. This will allow us to take a targeted approach to bring considerable opportunities for growth in the future...
>
> We expect to launch free Voicemail services to all our customers in the new calendar year and to roll out number portability to all new customers, thereby eliminating the last major obstacle to conversion to our service for many would-be customers. (CEO Louis Brunel, Videotron, Annual Report 1995)

NINE

Netcom Systems: Developing a Corporate Strategy

Starting in 1993, NetCom Systems AB was a wholly owned subsidiary of Industriförvaltnings AB Kinnevik. In 1996, the shares of NetCom were distributed to Kinnevik's shareholders. NetCom Systems AB, along with its subsidiaries and associated companies in Sweden, Norway, and Denmark, is the largest non-PTT provider of telecommunications services in Scandinavia. The Group is also engaged in cable TV operations. The Group operates within these areas (Annual Report, 1996):

- Public national and international telephony and data communications; in Sweden through its subsidiary Tele2 AB, and in Denmark and Norway, through other subsidiaries.
- Mobile GSM services; in Sweden, through the Comviq GSM AB subsidiary, and in Norway through an associated company.
- Cable TV services; in Sweden, through its subsidiary Kabelvision AB.

NetCom Systems AB reported an operating revenue of 2,957,733 SEK in 1996 (1,994,893 in 1995) and a net income of 247,659 SEK in 1996 (a net loss of 882,148 in 1995). The network building has primarily been financed through capital from the parent company, Kinnevik. In 1994, an international banking group including Westdeutsche Landesbank (Annual Report, 1993) was also engaged in the financing. Tables 9.1 and 9.2 provide further financial information.

Kinnevik's telecommunications ventures began with mobile telephony (Annual Report, 1994). Miscellaneous research had been carried out since 1976 within Miltolpe Group Inc., a company in the USA having shareholders in common with

Table 9.1 Consolidated statements of operations for NetCom Systems AB, SEK thousands (source: Annual Report, 1996).

	1996	*1995*
Operating revenue	2,957,733	1,994,893
Operating expense	(2,313,605)	(2,425,860)
Operating income (loss) before	644,128	(430,967)
Depreciation and amortization	(397,401)	(297,109)
Operating income (loss) after depreciation and amortization	246,727	(728,076)
Shares of income (loss) of associated companies	117,413	(400,104)
Financial items		
– Interest revenue	61,534	156,874
– Interest expenses	(372,302)	(464,643)
– Exchange rate difference	2,500	(535)
– Other financial items	(26,545)	(19,599)
Financial items, net	(334,813)	(327,903)
Income (loss) after financial items	29,327	(1,456,083)
Appropriations		
– Group contribution	–	381,250
– Shareholders' contribution	–	73
Income (loss) before taxes	29,327	(1,074,760)
Benefit for income taxes	241,898	210,624
Minority interest	(23,566)	(18,012)
Net income (loss)	247,659	(882,148)

Kinnevik. In 1979, these shareholders founded Millicom Inc. in the USA with the purpose of speeding up the pace of development there. Also in 1979, Kinnevik acquired, on an American initiative, a small mobile telecommunications operator in Sweden with a manual system to serve as a basis for the drawn-out legal and political process that would be needed to establish the principle of competition on equal terms with the

Table 9.2 Consolidated balance sheets for NetCom Systems AB, SEK thousands (source: Annual Report, 1996).

	Year ended 31 Dec.	
	1996	*1995*
Current assets		
– Cash	220,233	24,448
– Receivables from Kinnevik Group	226,986	453,700
– Accounts receivable	445,078	268,672
– Prepaid expenses and accrued revenue	194,175	239,209
– Other current receivables	37,342	59,093
– Materials and supplies	10,906	5,162
Total current assets	1,134,720	1,050,284
Fixed assets		
– Shares in associated companies	120,586	142,020
– Shares and capital interests	553	553
– Receivables from Kinnevik Group	–	326,420
– Other long-term receivables	14,178	137,457
– Deferred tax assets	448,977	247,079
– Intangible assets	1,888,542	718,989
– Property, plant, equipment	3,879,860	2,208,042
Total fixed assets	6,392,696	3,780,560
Total assets	7,527,416	4,830,844
Current liabilities		
– Interest-free	1,110,661	726,280
– Interest-bearing	69,545	2,416,097
Total current liabilities	1,180,206	3,142,377
Long-term liabilities		
– Interest free	9,555	8,858
– Interest-bearing		
Liabilities to Kinnevik Group	–	90,633
Bank credit facility	5,994	95,706
Convertible debenture loan	647,007	–
Long-term borrowings, external	3,406,215	2,430,283
– Total interest-bearing	4,059,216	2,616,622
Total long-term liabilities	4,068,771	2,625,480
Minority interests	2,339	(26,613)

Table 9.2 (*continued*)

	Year ended 31 Dec.	
	1996	*1995*
Shareholders' equity		
– Share capital	441,474	100
– Restricted reserves	2,825,411	(195,926)
– Total restricted equity	3,266,885	(195,826)
– Retained earnings	(1,238,444)	167,574
– Net income (loss)	247,659	(882,148)
– Total non-restricted equity	(990,785)	(714,574)
– Total shareholders' equity	2,276,100	(910,400)
Total liabilities and equity	7,527,416	4,830,844

Swedish telecommunications monopoly. This company was to become Comvik.

Parallel with the mobile telephony activities, preparations were started within traditional telecommunications for voice and data in the 1980s. A gateway for data traffic was opened in 1986, and in 1989 an agreement was concluded with the Swedish State Rail Administration for joint investment in a fiber optic network. The intention was to connect all large population centers in the country.

Tele2 was formed in the following year. When the telecommunications market was further deregulated in 1993, Tele2 was able to act fast and reached second place after Telia.

Even though the dominating company, Telia, is active in all segments of the Swedish market, competitive conditions vary widely from one segment to another. The variation is caused by such factors as market growth, technical advances, or historic tariff structures. In 1993, Kinnevik therefore opted to start individual companies in each segment and to adapt their strategies and allocate their resources on the basis of their short-term potential. Tele2, Comviq GSM, and Kabelvision thus brought separate brand names to their respective markets.

At the same time, the telecommunications business area of Kinnevik became an operational unit, led by NetCom Systems AB, instead of a project organization as before.

> NetCom Systems AB is a young company in the telecommunications industry. At the same time we have much experience of breaking into monopoly markets. This happened in Sweden which is one of the world's most deregulated markets...
>
> This means that NetCom is well prepared to exploit the opportunities that deregulations and new technology creates on the Nordic markets. (Managing Director Håkan Ledin, NetCom Systems AB, Annual Report, 1995).

A number of changes to the Group structure were initiated in 1996 when minority interests in subsidiaries were exchanged for shares in the parent company. As a result, NetCom Systems' Swedish telephony operations became wholly owned, which increased the opportunities to integrate within the company and to exploit synergies. The ambition was to consolidate and develop its position as the largest privately owned provider of telecom services in Scandinavia.

> NetCom Systems has moved from being a 'telecoms holding company' to being an operating group. The work of integrating operations in order to obtain synergies, both towards the market and within the organization, can now begin...
>
> NetCom's technical platform, which allows us to market seamlessly between the companies, is now in place. Examples of the services we will be offering include combined services with mobile/wired telephony and telephony/ broadband connections to the Internet using the cable television network (as access to the end-users). These services will allow us to break into areas where Telia has managed to keep its monopoly. (CEO and President Anders Björkman, NetCom Systems AB, Annual Report, 1996)

The interest of the individual companies in similar customer groups makes possible the marketing of a broader range of services to customers in the existing customer base. In turn, this leads to a need for co-ordination of corporate activities pertaining to administration and sales. The market organization of NetCom Systems has, hence, been divided into a "company market" department and a "mass market" department.

Furthermore, when it comes to technical aspects, it is a possibility to integrate the stationary, mobile, and cable TV networks and, for example, to use fewer control stations to supervise all three networks.

TELE2: CHALLENGING THE SWEDISH DOMINANT

Tele2 AB is a wholly owned subsidiary of NetCom Systems AB. With 22 percent of the Swedish market for international telephony and 6 percent of the market for national long-distance traffic (Annual Report, 1995) Tele2 is the largest private telephony operator in Sweden. The company also offers services for communications networks, such as Internet, X.25, and Lan, and also leased lines.

There are business areas for international and national telephony, business network services, and the Internet. The major customer groups consist of companies and residentials for standard telephony and companies interested in adapted communications solutions. Table 9.3 shows financial figures.

International and national telephony expanded in volume from 1995 to 1996. Because of a dramatic decrease in market prices, Tele2's volume expansion in international telephony was not/ accompanied by a parallel revenue growth. A constant gross

Table 9.3 Financial information for Tele2 AB, MSEK millions (source: Annual Reports).

	1996	*1995*	*1994*	*1993*	*1992*
Total sales	1,383	937	497	183	12
Operating expenses	1,105	735			
Income before depreciation	278	202			
Income after depreciation	145	101	(28)	(142)	(19)
Income after financial net	103	45			
Capital expenditure (machinery, equipment, infrastructure)	420	253	151		
Gross margin	20%	22%			

margin was kept by means of higher productivity and through lower settlement rates with foreign operators.

Internet services have become Tele2's second largest revenue source after international telephony. Volume in minutes increased by more than 1,000 percent during 1996. These services comprise dial-up modem connections for both companies and private consumers and fixed Internet connections for companies.

In 1995, Tele2 had two public exchanges, one outside Stockholm and one in Gothenburg. The fibre optic network is connected to Telia's local access network throughout the whole country. Traffic to Asia and Australia is controlled via a satellite. During the summer of 1995, Tele2 and US Sprint opened up the world's first 34 Mbits Internet link over the Atlantic, which increases the speed of transmission significantly.

Establishment in the Swedish market

Tele2 has its roots in Comvik Skyport AB, which was established in 1986. The original services comprised data transmission via satellite and the offering of equipment for the transmission of TV programs. In 1989, a long-term contract was signed with the State Rail Administration in order to build a common fibre optic network that uses the established railway network.

In 1990, the company name was changed to Tele2, and Cable & Wireless of England bought 39.9 percent of the company. Tele2 started to offer data network services addressed to companies and, later, also offered leased lines for national and international telephony. Broadscale marketing of public telephony services was initiated in March 1993, when companies and private consumers were offered cheap international calls by using the prefix 007. The corresponding service for national calls was introduced on 1 October 1994.

Tele2 has focused on building up its own infrastructure in order to be able to control its network costs. Having its own network all the way to the customer allows the company to compete not only with call rates but also with the fixed cost: the subscription charge.

Tele2's own sea cables to Denmark, Finland, and Latvia and its satellite connections from Kista to Asia make the company independent of Telia for its international traffic.

The agreement with the State Rail Administration covers joint investments in a fiber optic network that is intended to connect all large population centers in the country. Besides this, Tele2 has signed a contract with Svenska Kraftnät (Swedish Power Networks) making it possible to install fiber optic cables in the power network. In 1997, the major cities of Stockholm, Gothenburg, and Malmoe were linked together.

The State Railway Administration's network and Svenska Kraftnät are complemented in a number of large, industry-intensive areas with many potential customers by Tele2's own local fiber loops and microwave links. These are used to connect a number of corporate customers directly.

An interconnection contract with Telia makes it possible to penetrate residentials and small companies. These customer types are normally not profitable enough to be directly connected to the other networks. The price level stipulated in Tele2's contract with Telia implies, however, difficulties for Tele2 in trying to expand its national telephony.

The Telia relationship is regarded as one of the major challenges for the development of Tele2. On a number of occasions the company has charged Telia with abusing its dominant position:

> The response to the complaints has been that Telia has reduced its prices to customers for long-distance and international calls without any effect on the shared-traffic charge. For Swedish consumers,Telia's raising of local tariffs and subscription fees has meant that customers who make long-distance and international calls are the only ones benefiting from competition. It would be possible to change this situation by splitting Telia into a network company and a marketing company, totally independent of each other. (Tele2 AB, Annual Report, 1995)

Thus, the company's view is that competition in national telephony could be intensified if the Telia corporation was to be divided into one network company and one operational company which compete with other operators in the market.

Tele2 has an explicit ambition of expanding its infrastructure and/ seeking alternative ways to reach customers in order to control costs and to reduce its dependence on Telia. One alternative way of accessing the end-users is through the cable TV network controlled by the group. Exploitation of satellite technology is another option.

A further example of this is a cooperation agreement that was signed in the autumn of 1995 with Linköping Technical Public Utilities Service for the use of its 70-kilometer fiber optic network. This makes it possible to connect customers directly to the Tele2 network.

The existing access network which has been built up during many decades and financed through public taxes, and which is now controlled by Telia, is enormously costly to duplicate. It's a national scandal that parallel access networks have to be built up...

Even though we suggest that there should be only one stationary network company, the bargaining power of operators will secure cost efficiency of the network company. The very limited possibilities to reach end-users is the major reason for the high subscription rates in Sweden today, compared to other countries. (Market Director Pelle Hjortblad, Tele2, 29 October, 1997)

Initial strategy in the Swedish market

Tele2 views itself as an alternative to Telia and, accordingly, offers a broad range of telecommunications and data communications services, ranging from standardized telephony for small companies and private consumers to adapted communications solutions for large companies. The intention is to keep its position as Sweden's largest private telecommunications operator, and as a data communications actor:

We have the advantage of having operated in a deregulating market for some years.The new players are less well established. However, several of them will undoubtedly increase their efforts on the Swedish market. Niche operators have played a relatively small role in the market and will probably play an even smaller role in the future. (Annual Report, 1996)

The goals of the company also include preserving its position as a leading Internet supplier. Tele2's market strategy for the next few years comprises these elements:

- To continue to be the price leader.
- To offer high-quality services and the best customer service in the market.
- To offer a sufficiently broad product range to be able to provide a genuine alternative to Telia. (Services that are too complex are avoided, as they demand large personnel resources and, thus, are too costly.)
- To achieve cost effectiveness by having its own infrastructure.
- To extend co-operation with other NetCom Systems companies in order to achieve synergies.

The primary way of improving efficiency is to try to increase the volume of traffic for each employee. This makes it possible to expand total sales with limited personnel.

> We try to be a customer-driven company and offer communications capacity, low prices and service quality. We start out from customers' communications needs and then form the offerings. Our objective is that around 80% of the customers shall perceive Tele2 as the most interesting supplier. A higher percentage requires too many marketing resources...
>
> It is now crucial to increase revenues per customer, not to get new ones. Those customers with enough volume demands are being directly connected to our network. (Market Director Pelle Hjortblad, Tele2, 29 October, 1997)

In 1996 Tele2 operated throughout Sweden. Telephony services are offered by Tele2's own personnel for the major company accounts and through external channels, such as telemarketing, for small companies. These and private consumers are being penetrated through direct mail, TV and radio commercials, and outdoor advertising.

The entire product range is offered through direct access to medium-sized and large corporate customers. Corporate customers are offered specialty telephony services, such as freephone, virtual networks, combination services and directory assistance.

The highest percentage of its international traffic is to destinations to which Tele2 has its own marine cable links, including Denmark, Finland, and Latvia. Because of Sweden's geographical location, Tele2 has the ambition of acting as a hub for telecommunications and data traffic between East and the West.

COMVIQ GSM: BUILDING A MARKET POSITION IN MOBILE TELEPHONY

Comviq GSM AB is a wholly owned subsidiary of Netcom Systems AB. Comviq GSM possesses a nationwide network and competes with Europolitan and Telia in the Swedish GSM market. Comviq penetrates both the consumer market and the company market.

Comviq considers its main competitive advantages to be low prices, good indoor coverage, and good customer service, as well as its ability to respond quickly to new customer demands and market conditions.

The company has roaming agreements with 63 operators in 40 countries (Annual Report, 1996), which means that a customer can use a Comviq subscription to make calls in all of these countries. Additional agreements are continuously being added as the GSM system is being extended throughout the world.

Table 9.4 provides financial figures for Comviq GSM AB.

Table 9.4 Financial information for Comviq GSM AB, MSEK (source: Annual Reports).

	1996	*1995*	*1994*	*1993*	*1992*
Total sales	1,584	1.088	449	107	103
Operating expenses	1,154				
Income after depreciation	216	(63)	(140)	15	(14)
Capital expenditure	492	675			
Total number of subscriptions, 31 Dec., thousands	466	422	135	20	

Establishment in the Swedish market

During the 1960s and 1970s a number of companies operated mobile telephony networks in Sweden, usually as complements to original businesses. The majority of the networks were local, but some covered relatively large parts of the country. Mainly because of financial problems (Mölleryd, 1997) most of the pioneering operators disappeared, and in 1980 Företagstelefon AB ("Company Telephone") was the only private mobile telephony operator in Sweden. The mobile telephony operations of Företagstelefon were started by a watch retailer in the city of Jönköping in southern Sweden in the middle of the 1960s.

Technology development and changing political attitudes towards deregulation at the end of the 1970s created opportunities for new establishments in the Swedish telecommunications market. Hence, Kinnevik acquired Företagstelefon in 1979 and renamed it Comvik.

Based on the frequencies of Företagstelefon, Comvik made plans for an analog NMT system for mobile telephony; however, Televerket did not accept Comvik's application for a licence. Comvik appealed the decision and finally got permission to operate only manual radio exchanges and to use the public telephony network.

Comvik then turned to the government in order to be able to connect its automatic exchanges with the public network. After some discussions, Comvik was granted permission, although the government, in principle, considered voice communications through the public network to be a part of Televerket's monopoly. One reason for this was that Comvik had only 2,000 subscribers and, according to its forecast, there would be only 6,000 in 1990 (Mölleryd, 1997). Comvik received permission to engage 20,000 subscribers.

Thus, Comvik's system was operational in 1981. In 1982, Comvik started to engage its own sales personnel in order to capture subscribers for the network. With the help of governmental decisions, Comvik was permitted to use a growing number of radio frequencies during the subsequent years.

In 1988, Comvik received a licence to operate a network based on the new GSM technology. This network was opened in September 1992, and Comvik was renamed Comviq GSM. (Telia Mobitel and Nordic Tel also received licences.)

The initial business mission of Comviq GSM meant that the company would be offering mobile telephony services to companies and private consumers in Sweden (Annual Report, 1993). The company had taken more than 50 percent of the share of the Swedish market in 1993. The concentration on the emerging GSM technology implied that the analog system of Comviq was being closed and that the customers were being offered the opportunity to change over to Comviq's GSM service.

In 1996, the network had the geographical coverage demanded by the National Post and Telecom Agency. This requirement was that all European highways and cities with more than 10,000 inhabitants were covered. Most of the investment in infrastructure had been made. Part of the infrastructure consists of the State Rail Administration's fiber optic network. This connects all major urban areas in the country. Comviq's base stations are linked to the fiber optic network using the company's own radio link equipment. In 1995, Comviq GSM invested 556 million SEK in its infrastructure (382 million SEK in 1994).

Initial strategy in the Swedish market

In addition to regular voice telephony, Comviq offers its customers a number of complementary services in order to meet customer requirements. Voice mail, number presentation, and the possibility of sending short written messages using the display of the telephone are examples of such additional services. Specific corporate customer services and the corporate sales department will be extended. Low prices and high call quality in urban areas are being underscored as major competitive advantages for all customer types.

More intensive cultivation of the company market in 1995 than in earlier years led to a number of large-customer contracts, including agreements with Norrköping Municipality, IBM, Swedish

State Railways, the City of Stockholm, and the Stockholm County Council (Annual Report, 1995). A number of services and forms of subscriptions were developed to meet the needs of the company market.

The number of customer corporations with volume agreements increased by 30 percent in 1996, mainly due to adapted campaigns, an extended sales force, and adapted subscription types.

Close cooperation with leading retail chains is regarded as the core of Comviq's marketing strategy (Annual Report, 1995). The company offers payments to dealers for each new Comviq customer plus a payment related to the customer's use of the service. The company trains dealers' sales personnel and in other ways provides direct sales and marketing support designed to promote greater sales. In the company market, a Comviq salesforce sells to corporations directly. TV commercials are used to a large extent in order to make the brand well known.

The creation of relationships with retail chains is a crucial way of reaching consumers. Until August 1993, all GSM subscriptions and telephones were sold by traditional mobile telephony retailers. After that, radio and TV outlets also started to offer mobile telephony. Mobile telephones were often used as loss leaders in marketing activities directed towards potential new subscribers. Prices could be lowered thanks to lower costs at the manufacturing stage and stepped- up support from systems suppliers (Annual Report, 1994). For example, in 1995 prices of phone handsets fell by approximately 25 percent.

Comviq has chosen not to set up its own stores and instead works solely with independent dealers. These are divided into three groups (Annual Report, 1996): Independent retailers and buying chains; Mobile telephony chains; Radio and television outlets.

The first two groups have a strong hold on the company market, largely due to their earlier ties with the NMT system, the precursor of GSM in Scandinavia, while Radio and television outlets are strong in the consumer market. In 1992, specialty stores accounted for nearly 80 percent of the total market, and radio and television outlets accounted for approximately 20 percent. These figures were almost completely reversed in 1996.

PART IV

Discussion of Empirical Findings: Patterns and Processes

INTRODUCTION

We are now able to discuss the empirical information as regards strategies of telecommunications operators, including information both from the overview of competition in the UK and Swedish markets and from the cases, which, to some extent, also contain information on expansion into other European markets.

Based on empirical classifications previously presented in this book, the discussion in this part leads to the identification of patterns following the two models of business strategy on emerging markets: the market establishment model and the strategic states model (Chapter 10).

Moreover, the notion that strategy frequently does not develop in a straight line, but through a series of complicated variations, requires that relevant processes are highlighted (Chapter 11). Processes of corporate restructuring and integration consistent with corporate strategy intentions are, thus, exemplified based on case study information.

Chapter 12 focuses on the applicability of the empirical findings to emerging markets other than telecommunications markets for operators in the UK and Sweden, such as markets for other growing industries or geographical regions. As every company needs an explicitly planned strategy in order to survive under profitability constraints in competitive environments, some of the advice given here is valid for strategy in emerging markets in general.

TEN

Business Strategies of Telecom Operators in the UK and Sweden

This chapter relates descriptive findings generated through the industry overview and the case studies to the models of business strategy in emerging markets: the market establishment model and the strategic states model. Thus, empirical patterns are identified following the framework of the models.

PATTERNS IN RELATION TO THE MARKET ESTABLISHMENT MODEL

In the market establishment model of this book (Figure 10.1), perceptions of entry barriers imply varying assessments of business opportunities. These assessments influence the choice of entry strategy. On the contrary, the process of strategy implementation may result in changing perceptions due to

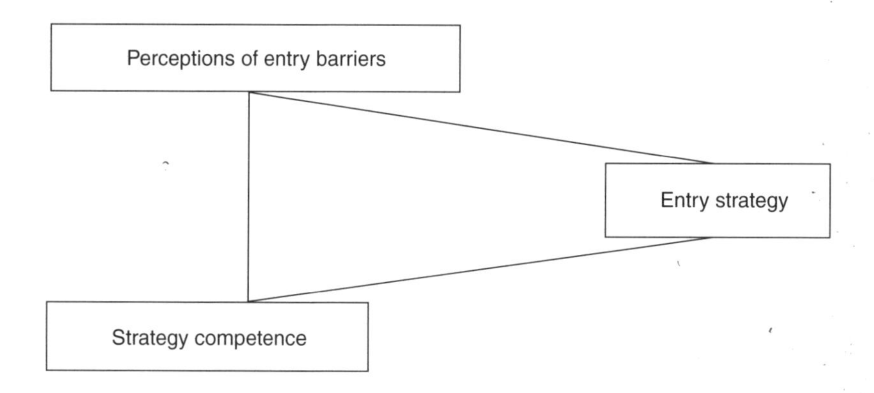

Figure 10.1 The market establishment model.

learning effects. For example, the assessment of a specific phenomenon in the environment may differ from one implementation phase to another. Moreover, both perceptions and experiences gained during the strategy process are interrelated with comprehension of the need for competence in order to become firmly established in the market in question.

The following discussion initially pays attention to descriptive patterns valid for each of the components of the model and then continues with analyses of company cases in order to define relationships between the components of the model specific to individual company processes of market establishment.

Patterns of perceptions of entry barriers

This book focuses on two important barriers to entry: the perception of deregulation and the capital needed for infrastructure investments. Such barriers to entry are perceived differently from company to company and from one individual to another. Perceptions basically are subjective interpretations of available information, including a certain degree of uncertainty, and this results in different assessments of the importance of barriers.

Here, perceptions of entry barriers primarily concern perceptions of environmental states consisting of so-called "state uncertainty", manifested by, for example, an inability to predict the future behavior of a key competitor. Since complete information on the present situation as regards entry barriers and on expected changes is seldom available, each industry actor has to rely on limited information and can enjoy only varying degrees of confidence in their assessments. This is also the case if there is too much information available.

Although the process of deregulation started in the late 1970s in the UK and Sweden, regulatory hurdles still exist in both countries. The licence requirement for operators of telecommunications is one such barrier to entering the markets. In both countries, there is much debate related to the licences.The debate mainly concerns the roles of, for example, the regulators (Oftel and The National Post and Telecom Agency) and the dominating operators (BT and

Telia). More specifically, the issues include, among other things, social obligations and the treatment of interconnection fees.

Quoted previously, this is just one example of how the role of the regulator in the UK is perceived and how it relates to the market dominance of BT:

> One obstacle in our path continues to be over-complex and increasingly unworkable regulation. This now threatens to impede the growth of competition, with customers as the losers. We are fast approaching regulatory failure, which only fundamental reform can resolve. Our recommendations for this reform require increased powers for the Director General of Oftel to curb anti-competitive behaviour and the removal of artificial barriers to competition. (CEO Peter Howell-Davies, Mercury Communications, Annual Report, 1995)

However, Oftel is, to a large extent, concerned with BT's behavior in the process of deregulating the UK market:

> It is important that BT should recognise that its own behaviour will be a major determinant of whether the UK market can be considered one of effective competition. At the moment BT too often appears reluctant or unwilling to address with Oftel and the industry the issues of clear regulatory concern. It needs to have a more positive approach to improving the competitive framework in the UK. (Oftel Statement, 1995)

BT, on the other hand, perceives the deregulation process as uncertain and the regulatory environment as hostile and unpredictable:

> In the UK, we face significant competition from the cable companies and other licenced operators. Competition is flourishing in the local, long-distance, international and mobile markets. But, provided that it is fair and equitable competition, we welcome it as in the best interest of customers and the industry as a whole.
>
> When I wrote to you last year, I suggested that in the UK we suffered from "the paradox of greater competition accompanied by greater regulation." This year, I believe that our relationship with our regulator, Oftel, has undergone a step change. Although BT accepted Oftel's proposal for accounting separation, which requires us to publish more detailed cost and pricing information, we were not able to accept the regulator's proposals on number portability. Consequently, Oftel announced that it would be referring the issue to the Monopolies and Mergers Commission. This is the first time BT has experienced such referral, but we are convinced that we are right to reject Oftel's proposals.
>
> This reference is just one example of how Oftel requires us, in effect, to subsidise competitors, many of whom are large, rich American companies which, ironically, enjoy monopoly or near-monopoly status in their home markets." (Chairman Sir Ian Vallance, British Telecommunications, Annual Report, 1995)

In Sweden, Tele2 has signed an interconnect contract with Telia, the market dominant, which allows it to use Telia's stationary local access network. This means that Tele2 is able to penetrate even residentials and small companies, as these customers are normally not profitable enough to be directly connected to Tele2's own network. The price level controls in the contract with Telia imply, however, difficulties for Tele2 in its efforts to try to expand its national telephony operations. In fact, the Telia relationship is considered to be one of the major challenges for Tele2. The company has, on a great number of occasions, charged Telia with abusing its dominant position:

> The response to the complaints has been that Telia has reduced its prices to customers for long-distance and international calls without any effect on the shared-traffic charge. For Swedish consumers, Telia's raising of local tariffs and subscription fees has meant that customers who make long-distance and international calls are the only ones benefiting from competition. It would be possible to change this situation by splitting Telia into a network company and an operating marketing company, totally independent of each other. (Tele2 AB, Annual Report, 1995)

These examples clearly illustrate how the process of deregulation is perceived differently from one company to another, resulting in, for example, different opinions of the roles of the major actors and conflicts of various types.

Based on an analysis of the liberalisation of telecommunications in Sweden, Karlsson (1998) concludes that the common feature of all conflicts was that they were initiated as a result of interconnection attempts at the boundary of the traditional telecommunication system. In fact, new technology development and technological convergence are generally the major driving forces behind interconnection efforts.

Karlsson further argues that conflict situations frequently involve colliding sociotechnical cultures. The business culture of new entrants confronts the "monopoly culture" of previous monopoly operators, resulting in the establishment of a conflict zone. In the Swedish case, each and every one of the conflicts have been transferred to the political arena at one time or another for settlement. Hence, emerging conflicts among major actors have been a decisive input into the policy-making process.

Besides the licence requirement, any company that has a desire to enter the industry will face the high need for capital, including the arrangement of infrastructure. No matter if the operator tries to build its own network or whether it rents capacity, the capital demand is high. Networks based on fibre optic technology have become a common substitute for networks of traditional copper cable. As direct connection to these new networks requires the penetration of high volume customers in order to be profitable, large international companies are the main target group for direct fibre optic connections.

Local connections are frequently made possible through interconnections to existing local access networks based on copper cable. Together with the evolution of mobile telephony, however, radio technology is becoming a realistic alternative in local loops.

Tele2 of Sweden has an explicit ambition of expanding its infrastructure and seeking alternative ways of reaching customers in order to be able to control network costs and to reduce its dependence on Telia. One alternative way of accessing residential end- users is through the cable TV network controlled by the group, while exploitation of satellite technology is another option:

> The existing access network which has been built up during several decades and financed through public taxes, and which is now controlled by Telia, is enormously costly to duplicate. It's a national scandal that parallel access networks have to be built up.
>
> Even though we suggest that there should be only one stationary network company, the bargaining power of operators will secure cost efficiency of the network company. The very limited possibilities to reach end users is the major reason for the high subscription rates in Sweden today. (Market Director Pelle Hjortblad, Tele2, 29 October, 1997)

COLT is a recently established operator in the UK that offers services over its own fibre optic network. The network construction, however, puts a heavy strain on its liquidity and capital reserves. The costs associated with the initial installation and expansion of COLT's networks in London and Frankfurt, including development, installation, and early operating expenses, have been significant and are expected to result in increasing negative cash flows:

> Negative cash flow is expected to continue in each of the company's markets until an adequate customer base and related revenue stream have been established. COLT believes its operating losses and negative cash flow will increase with the continued expansion of its networks. (Chief Financial Officer Lawrence M. Ingeneri, COLT Telecom plc, Annual Report, 1996)

The company believes that its existing resources, together with internally generated funds and loans expected to be available under its bank facility, will provide sufficient funds for the company to expand its business as planned.

Patterns of strategy competence

As I view it, the likelihood of becoming firmly established in a market depends, to large extent, on the relevant strategy competence of the company in question. That is, the higher the competence, the better the likelihood of becoming firmly established in the market. This requires acquisition of knowledge on strategy formulation and implementation and relevant knowledge on the products, services, and markets involved.

Furthermore, I argue that the strategy competence of a certain company is due to the position of its businesses in relation to the original core business of the corporation (that is, "relatedness"). This means that efforts to establish a business that may be considered a part of the original corporate core business are, hypothetically, the most likely to be successful. The opposite is valid for a business unit that is less related to the core of the corporation.

In general, at least three tests can be applied to identify the core competencies of a company (Prahalad and Hamel, 1990). First, core competence provides potential access to a wide variety of markets. Second, a core competence should make a significant contribution to the perceived customer benefits of the end product. Finally, a core competence is the kind of competence that is difficult for competitors to imitate. Generally, the core business of a corporation is the center of core competence development, although the competence helps in penetrating markets valid for other businesses. This requires a smooth distribution of skills among involved businesses relying on the core competence.

The capability of constructing networks for telephony in order to be able to penetrate the desired target customers and market segments interested in telephony services may be defined as the core competence of any telecommunications operator. In some cases, the construction of local access networks is the relevant core competence, while competencies needed to construct networks relying on fiber optic technology or radio technology may constitute the core in other cases. Such competencies, obviously, make possible the penetration of a variety of market segments, and network capability and quality generally make an important contribution to perceived customer benefits as any network is the basis for the offering of services. Finally, intricate question of reaching end-users through local access networks demonstrates the difficulties inherent in imitating core competencies.

The definition of core competence given above implies that the core business of a telecommunications operator would be the unit that is engaged in the construction of telephony networks. However, as core competencies develop over time, telecommunications operations of any corporation may be more or less related to the original corporate core business, which is based on knowledge of telephony networks or on any other skill.

In the UK market there are companies with telecommunications operations that do not belong to the original corporate core businesses (Figure 10.2). For example, electricity companies and cable TV companies based in the UK have found their way into the telecommunications industry using networks originally designed for other purposes. Cable TV companies exploit their fibre optic networks for local access and often provide telephony and related services in addition to what they have traditionally offered. This organic type of strategy development has not, at the time of writing, yet occurred in Sweden, although

Relatedness	Market experience	
	Early entrants	Late entrants
Telecom operations are part of the original corporate core business	The UK -British Telecom -Kingston -Mercury -Mercury PC -Vodafone Sweden -Telia -Tele2 -Comviq	The UK - COLT -MFS -Scottish Telecom -Telstra -Worldcom Sweden -Telenordia -Global One -MFS -Telecom Finland -Europolitan -Tele8
Telecom operations are not part of the original corporate core business		The UK - Atlantic -Orange -Racal -Scottish Hydro-Electric -Videotron Sweden -Dotcom

Figure 10.2 Classification of UK and Swedish operators according to relatedness and market experience.

there are examples of co-operations between telecom operators and companies possessing other networks. Thus, in the Swedish market almost every operator belongs to a genuine telecommunications corporation with a high degree of relatedness among businesses. Two international alliances are also present in this market: Global One (representing Deutsche Telekom, France Telecom, and US Sprint) and Telenordia (BT, Tele Denmark, and Telenor of Norway).

Besides relatedness, strategy competence includes experience in the market. Familiarity with market conditions clearly strengthens the competence and the likelihood of finding a sustainable position in the market. Thus, increasing knowledge of products, services, and other market conditions gives extended market experience.

The patterns of market experience are basically of the same character in both countries. The dominants in the markets (BT and Telia) have their roots in the 19th century when national monopolies were formed with the help of governmental initiatives. In the UK, however, the Kingston operator survived as a municipally owned company. The first real challengers, Mercury and Tele2, to the dominants emerged in the 1980s and they belong to the group of early entrants as well. This is also valid for the first privately owned mobile telephony operators in both countries.

The late market entrants started to compete in the beginning of the 1990s in response to deregulation initiatives. This group comprises operators originating from the UK, Sweden or other countries. Moreover, the original corporate core businesses vary somewhat from one company to another.

The case of Videotron demonstrates the endeavors of a company with telecom operations that were less related to the original corporate core business and that had limited experience in the telephony market. Despite having a large portion of the London market, Videotron did not succeed in becoming firmly established in this local market.

Starting in 1991, regional licences were granted in the UK to cable TV companies and these were permitted to offer not only television services but also telephony across a local network. Videotron Holdings plc (later merged with Cable & Wireless Communications) became one of the largest integrated cable TV and telecommunications operators in London. The group was awarded licences to construct and operate broadband cable networks providing a range of television and telecommunications services within certain franchise and licence areas in London and Hampshire.

On 31 August 1996, Videotron had 134,975 households subscribing to its cable TV service, 123,133 residential telephony customers, and 18,654 business exchange lines connected. The network had reached 58 percent of the total franchise area that was required by its licence obligations.

Videotron's acquisition of contiguous franchise areas facilitated construction of a single integrated fibre optic network. This was supposed to lead to economies of scale and would allow

Videotron to offer businesses the option of networking their premises. Before the end of the decade, Videotron planned work investments involving over 1 billion pounds, besides the 400 million pounds that had already been spent.

Videotron's price advantage over BT was estimated to be 10 percent for a typical end- user, in part because residential customers were able to make free local voice calls to other Videotron customers. The same type of price advantage was also available for business customers.

The chief executive officer expressed his view on the need for market knowledge:

> Our strategy involves a number of initiatives. Firstly, we are focusing on sales and marketing, and have increased the size of our direct sales force. We are developing greater use of alternative sales techniques such as telemarketing and the use of external sales agencies. We have arranged to expand our presence in retail distribution outlets. For the first time we will mailshot and contact the existing base of homes twice in the forthcoming year whilst working hard to raise awareness of our services.
>
> We are also in the process of taking a fresh look at the segmentation of the residential market. This will allow us to take a targeted approach to bring considerable opportunities for growth in the future. (CEO Louis Brunel, Videotron, Annual Report, 1995)

Management of Videotron obviously tried to increase volumes once the decision to expand the network coverage was taken, and network investments were very costly. In turn, network expansion was a condition for acquiring the licence rights.

As the citation above indicates, Videotron lacked enough knowledge of customer needs and was not able to segment the market appropriately in order to be able to reach relevant customers efficiently. In my opinion, this illustrate a lack of sufficient strategy competence and, to a large extent, explains the high financial losses on the ordinary activities of Videotron in 1995 and 1996 (13,161 and 16,288 GBP thousands, respectively, compared to a total turnover of 57,490 and 81,757.)

Patterns of entry strategy

This book focuses on entry strategy in the sense of entry modes. A market entry mode is an institutional arrangement necessary for the entry of a company's products, technology, and human and financial capital into a market. I view three modes as different ways of organic development of sole ventures: exporting, assembly, local manufacturing; while four modes fall into the category of strategic alliances: licensing, contract manufacturing, management contracting, joint venture. In my opinion, market entry through an acquisition also represents a type of organic development, although not originally based on a sole venture. The main reason for this is that an acquisition is one way to carry out an establishment with the ambition of controlling the forthcoming process completely. An acquisition entry occurs when an existing competitor in an existing market is acquired by a company not previously established in that market. The acquirer should have the intention of using the acquired company as a base for expansion, and not just of holding it as a portfolio investment. Here, acquisitions also include mergers.

Figure 10.3 shows the modes initially applied by telecommunications operators entering the UK and Swedish markets.

Of course, the pattern in the figure is not static, which the following example illustrates. Cable & Wireless is based in the UK and has its roots in the Eastern Telegraph Company which was formed in 1872. This company linked London to Hong Kong by laying out and operating communications through the use of telegraph cables. The company got its present name in 1934.

A major ambition of the Cable & Wireless group is to be one of the leading international telecommunications groups in the world. In markets where the group has stationary line communications, the aim is to offer mobile communications, and vice versa. Where the group offers data services, it wants to add video, multimedia, and cable TV. In fact, Cable & Wireless considers itself to be a pioneer in offering not only domestic and international telephony but also entertainment services over a single network.

In October 1996, Cable & Wireless announced that it had reached an agreement to create a larger provider of integrated telecommunications, information, and entertainment services by forming Cable & Wireless Communications plc from a merger of

Entry modes	The UK market	The Swedish market
Organic development		
- Sole ventures		
* Home-based operators	BT, Kingston, Mercury, Racal, Scottish Hydro, Scottish Telecom, Vodafone, Cable&Wireless	Dotcom, Europolitan, Tele2, Telia
* Foreign-based operators	COLT, MFS, Orange, Telstra, Videotron, WorldCom	MFS, Telecom Finland Tele8
- Acquisitions	Atlantic	Comviq
Strategic alliances	Mercury PC	Global One, Telenordia

Figure 10.3 Initial modes of entry into the UK and Swedish markets.

the operations of Mercury Communications Ltd, Nynex CableComms Group plc, Nynex CableComms Inc., Bell Cablemedia plc, and Videotron Holdings plc. The new UK-based company would offer local, national and international services in voice and data, together with multi-channel television, Internet services, and, in due course, interactive multimedia. Thus, Cable & Wireless has shifted from the organic development of sole ventures to a merger mode.

Tele2's mode of entering the Swedish home market exemplifies another organic development of a sole venture. Tele2 has its roots in Comvik Skyport AB, which was established in 1986. The original services comprised data transmission via satellite and equipment for the transmission of TV programs. In 1989, a long-term contract was signed with the Swedish State Rail Administration for the purpose of building a common fibre optic network that would use the existing railway network.

In 1990, the company was renamed Tele2 and Cable & Wireless of the UK bought 39.9 percent of the company. Tele2 started out offering data network services addressed to companies and, later, also offered leased lines for national and international telephony. Broadscale marketing of public telephony services was initiated in 1993 when companies and private consumers were offered cheap international calls. One year later, national call facilities were also provided by Tele2.

COLT is a foreign-based operator competing in the UK that has also developed organically. The company was founded in 1992 by Fidelity Capital, a wholly owned subsidiary of Fidelity Investments, a mutual funds company, following its experience in co-founding Teleport Communications Boston. Founded in 1987, this is a local carrier in Massachusetts in the USA that provides services to business customers on its completely fiber optic network and that connects users to long- distance carriers or to other user locations.

COLT was formed to provide similar services to business users in London. Having been awarded an operator's licence in April 1993, COLT constructed an initial 15-kilometer fibre optic backbone in the heart of the City of London before launching services on its network.

Telecom Finland has also gone through organic development of sole ventures. In recent years, this operator has extended its presence in international markets; the main target markets are countries adjacent to Finland, such as northwest Russia, the Baltic countries and Sweden.

Telecom Finland entered the Swedish market in 1994. Here, the company primarily addresses advanced network services for data communications, such as multimedia and Internet connections, and telephony services to companies and organizations. In 1995, an operations and maintenance center and a customer support function were organized in Stockholm. That year, Telecom Finland intensified its focus on the expansion of group companies and associated companies established earlier. Its aim is to increase the direct export volume of domestically developed services to international customers:

> On the international side, in areas adjacent to Finland the focus was (in 1995) on the start-up and improved management of both fixed and mobile projects, while in more distant locations the activities centered around mobile projects. (CEO Aulis Salin, Telecom Finland, Annual Report, 1995)

Strategic alliances appear in both the UK and Sweden. For instance, a goal of Telecom Finland is to enhance the value of its international joint ventures by expanding both its networks and the business volumes carried over these networks. In fact, an objective is to increase sales of the international alliances so that they will account for one-third of net sales of the Telecom Finland group.

Other empirical findings show (Rutihinda, 1996) that the establishment of new and sole ventures is a favoured way of entering foreign emerging markets, particularly for high-technology firms. Successful acquisitions are frequently regarded as too demanding, as they require far-reaching screening efforts and organizational facilities, such as having to set up a special committee for evaluating the compatibility of the acquisition target with the parent company structure. Such requirements encourage foreign-based entrants to choose organic development of sole ventures which enable them to build facilities step by step and to extend their market experience. Moreover, organic development generally also enables the high-technology firm to exploit its technological capabilities in a controlled manner.

Thus, my findings clearly show that most telecommunications operators entering the markets in the UK and Sweden initially prefer the organic development of sole ventures, no matter whether the single operator is based in the local market or whether it is based in a foreign country. Although this is often a time-consuming entry mode, advantages such as the possibility of efficiently transferring technology know-how to the new ventures are obviously considered to be more important than gaining quick market shares through acquisitions or through forming alliances.

Processes of market establishment

Kingston of the UK and Comviq GSM of Sweden exemplify the emerging character of entry strategy. Kingston was established in the UK following an organic development of a sole venture, while Comviq GSM was originally based on an acquisition entry.

The market establishment processes of these companies (Figure 10.4) show that assessments of business opportunities based on perceptions of entry barriers decided which entry strategies were regarded as appropriate. On the other hand, the implementation of entry strategies clearly resulted in changing perceptions as the individuals involved and the companies learned about market conditions throughout the processes. Consequently perceptions of entry barriers changed as a result of experiences gained during the implementation processes.

Moreover, the two examples indicate that both perceptions of entry barriers and knowledge acquired during the strategy processes are strongly interrelated to comprehension of the need for competence in order to become firmly established in the markets in question.

In the early 1890s the city corporation of Kingston-upon-Hull made plans for a municipal telephony system to compete with the established Post Office and the privatly owned National Telephone Company (NTC) networks. During the 1890s, the Post Office was threatening the NTC by taking over their national trunk network. The Postmaster General of the UK then set about

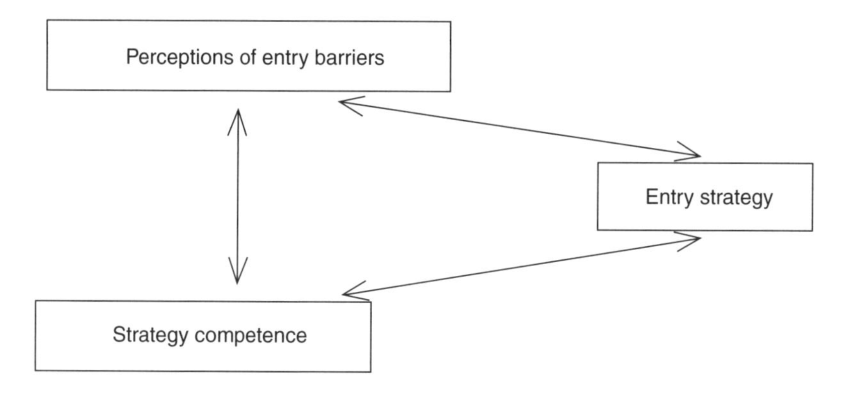

Figure 10.4 The market establishment processes of Kingston and Comviq GSM.

establishing a local Post Office network to take on the NTC in rapidly expanding urban areas. To quicken the pace of competition, he also allowed local municipalities to borrow money on the security of their general rates so that they could set up their own telephony networks under a Post Office licence. There was one condition: the Postmaster General retained the right to buy back the municipal telephony systems.

To attempt to establish a council-owned telephony company was a decision few local authorities were prepared to make. Out of 1334 local authorities, 55 expressed an interest. Some 13 applied for a licence and six actually set up service (Brighton, Glasgow, Hull, Portsmouth, Swansea, Tunbridge Wells). On 8 August 1902, Hull Corporation was granted its first licence.

In 1911, a decision was made that was to establish Hull's independence for years to come. The Postmaster General had taken steps to secure a UK monopoly of telephony services, buying out the NTC and a number of the local authority-owned services that were struggling as a result of poor planning or commercial failure, but he made Hull's bid for a new licence conditional on the purchase of the NTC network in the city. The council approved the purchase, and the sole municipally owned telephone company survived.

Comviq GSM has its roots in a company that also broke through barriers to enter the industry. During the 1960s and 1970s a number of companies operated mobile telephony networks in Sweden, usually as unrelated complements to original businesses. The majority of the networks were local, but some covered relatively large parts of the country. Mainly because of financial problems, most of the pioneering operators closed their telephony businesses, and in 1980 Företagstelefon AB ("Company Telephone") was the only private mobile telephony operator in Sweden.

Technological development and changing political attitudes towards deregulation at the end of the 1970s created opportunities for new establishments in the Swedish telecommunications market. Hence, Kinnevik observed what was going to happen and acquired Företagstelefon in 1979, renaming it Comvik.

Based on the frequencies of Företagstelefon, Comvik made plans for an analog NMT system for mobile telephony; however, Televerket (the Swedish public telecommunications administration) did not accept Comvik's application for a licence. Comvik appealed the decision and finally got permission to operate manual radio exchanges and to use the public telephony network.

As noted in the previous chapter, Comvik then turned to the government in order to be able to connect its automatic exchanges with the public network. After intensive discussions, Comvik was granted permission, although the government, in principle, considered voice communications through the public network to be a part of Televerket's monopoly. One reason for this was that Comvik had only around 2,000 subscribers and, according to its forecast, there would be only 6,000 in 1990. Comvik received permission to engage 20,000 subscribers.

Thus, Comvik's system was operational in 1981. In 1982, Comvik started to engage its own sales personnel in order to capture subscribers for the network. With the help of governmental decisions, the company was permitted to use a growing number of radio frequencies during the following years.

In 1988, Comvik received a licence to operate a network based on the new GSM technology. This network was opened in September 1992, and Comvik was renamed Comviq GSM.

The initial business mission of Comviq GSM was that the company should offer mobile telephony services to companies and private consumers in Sweden. In 1996, the network had the geographical coverage demanded by the National Post and Telecom Agency, and the majority of the investments in infrastructure had been made.

The establishment process of Mercury Communications was more responsive, in that the company reacted to new regulations. Therefore, it is appropriate to say that perceptions of entry barriers affected both evaluation of aspects related to strategy competence and the choice of entry strategy. However, even the case of Mercury shows a relationship between strategy competence and entry barriers (Figure 10.5).

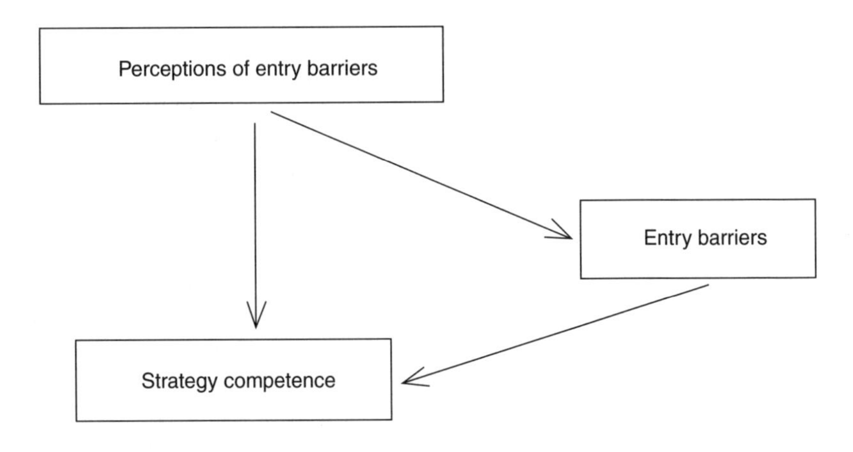

Figure 10.5 The market establishment process of Mercury Communications.

Mercury Communications was formed in 1981 as a partnership between BP, Cable & Wireless and Barclays Merchant Bank in order to compete with BT. However, by the end of 1984, Cable & Wireless had become the sole owner by acquiring all the shares.

As a result of the Telecommunications Act of 1981, Mercury alone was granted a provisional operating licence, which was extended to a 25-year Public Telecommunications Operators licence in 1984. This licence permits the operation of national fixed-link telephony lines.

In 1991, cable TV operators were licensed to offer not only television but also phone services across local networks. The response of Cable & Wireless to this was to buy 20 percent of the holding company for Bell Canada Enterprises, who operated a UK cable television and telecommunications franchise. In return, Bell Canada bought 20 per cent of Mercury shares. This strategic alliance offered mutual benefits: Mercury was able to bypass the use of BT lines for local calls, and Bell Canada was able to offer its customers cheaper long-distance and international calls. Mercury's strategy competence was thus affected in that the company portfolio was broadened.

PATTERNS IN RELATION TO THE STRATEGIC STATES MODEL

This section presents a discussion on the development of business strategies within the framework given by the strategic states model. Strategy clusters are presented and strategy development is exemplified by company cases. This leads to a discussion of competitive edges. In addition, performance implications of business strategy are noted.

Strategy clusters

The classification of operators competing in the UK and Sweden in accordance with the strategic states model generates four clusters (Figure 10.6) which will be discussed in what follows.

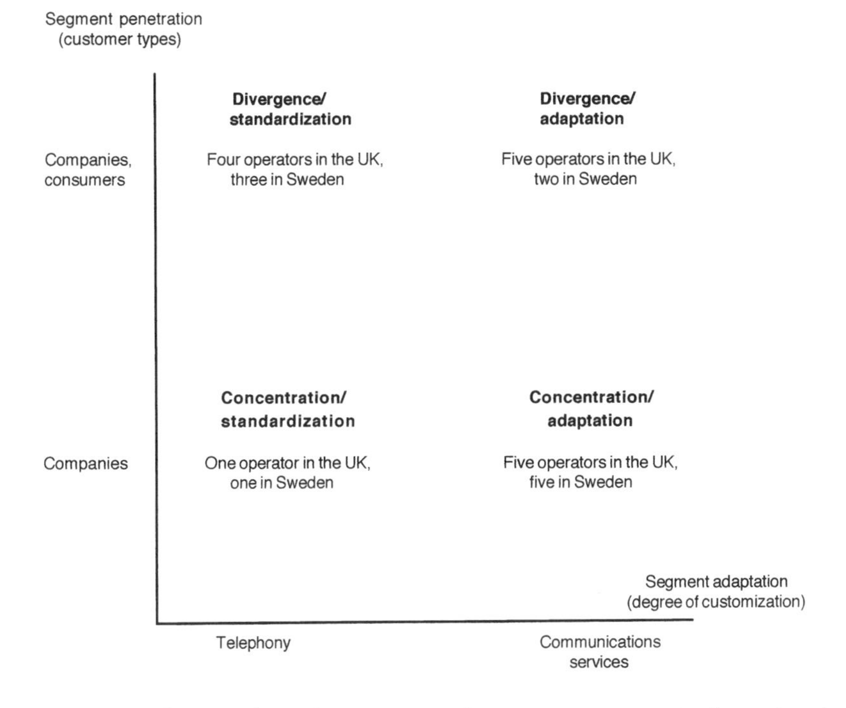

Figure 10.6 Number of operators in four strategic states in the UK and Sweden in 1996.

The concentration/standardization cluster

The strategic state that is characterized by concentration on the company segment and the offering of standardized telephony is not common in both the UK and Sweden. Worldcom, in principle, penetrates all major international industry sectors and company sizes in the UK and provides national and international calls through direct connections to its fiber optic cable network. In Sweden, Tele8 penetrates small and medium-sized companies and offers international traffic.

The divergence/standardization cluster

Mobile telephony operators in both countries are situated in the state of divergence (penetration of companies and private consumers) and standardization. Moreover, in Sweden the established mobile telephony operators co-operate with operators of stationary telephony in order to be able to offer more complete assortments. Comviq GSM, hence, cooperates with Tele2, while Europolitan/Global One and Telia Mobitel/Telia have become partners.

To give an example, as we have seen the initial business mission of Comviq GSM in 1992 was that the company should be offering mobile telephony services to companies and private consumers in Sweden. In addition to regular voice telephony, Comviq offered its customers a number of complementary services in order to meet customer requirements. Voice mail, number presentation, and the possibility of sending short written messages using the display of the telephone are examples of such additional services. In 1995, the company segment was more intensely cultivated directly by Comviq's own personnel, and this led to the development of a number of additional services and forms of subscriptions to meet the needs of the company market. The number of customer companies with volume agreements increased by 30 percent in 1996. Figure 10.7 illustrates Comviq's development of its business strategy.

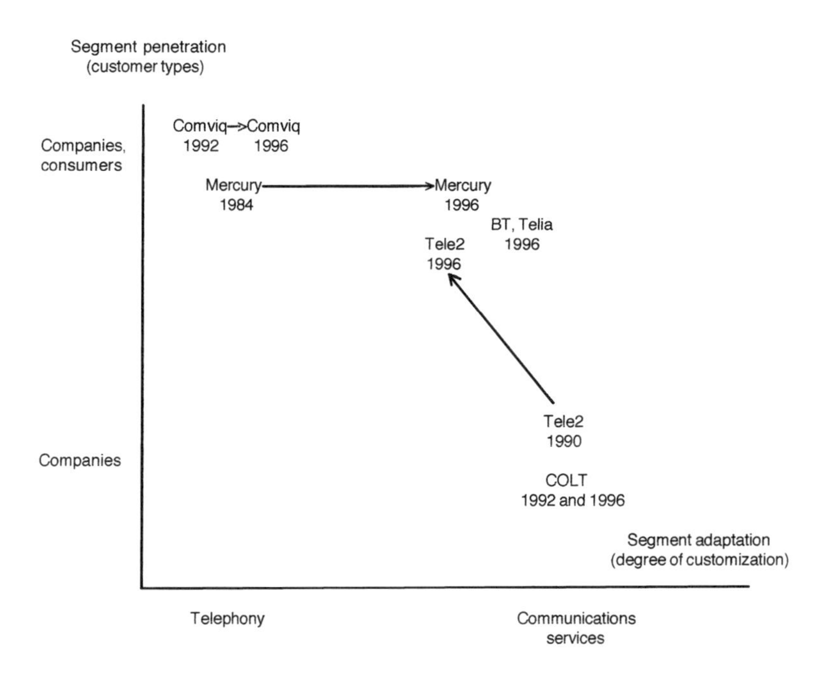

Figure 10.7 Business strategy development of companies operating telecommunications networks in the UK or Sweden.

The concentration/adaptation cluster

All of the UK and Swedish operators that are situated in a state of concentration on companies and on the offering of broader customized communications services, including telephony, entered the market late. This indicates that adaptation to certain company customers has become a popular initial business strategy among new entrants.

For example, COLT is an operator that was founded in 1992 and aims to be the preferred supplier to high-volume telecommunications users and carriers in major metropolitan areas, primarily in financial and business centers. The prime target is large financial, media, corporate, and governmental users and carriers, for whom reliable telecommunications are crucial. In 1996, London and Frankfurt were the main metropolitan areas on which COLT focused.

COLT offers a wide range of local telecommunications services and competes with incumbent public telephony operators by emphasizing high quality, integrated services (primarily over fibre optic digital networks) to meet the voice, data, and video transmission needs of its customers.

COLT's business strategy in the UK has been to focus on the London market by continuously evaluating and adapting to the telecommunications needs of its target users and to offer comprehensive services by building and operating a high-availability network through the use of state-of-the-art technology. COLT aims to provide the same mixture of service, cost-effectiveness, and flexibility that has been applied in London, to customers in Frankfurt and to the rest of the UK as it starts to roll its services out to other UK and European cities.

At the end of 1996, COLT's London network of 129,000 route kilometers provided switched and non-switched services to 537 directly connected customers in 583 buildings. The network embraced the major business concentrations of the City of London, the West End, Docklands, and Westminster.

In March 1996, COLT initiated service in Frankfurt. At the end of the year, this network covered 69,000 route meters and connected 90 buildings. Customers include banks, financial service providers, and carriers such as AT&T and Unisource. Seven months later, COLT was granted additional infrastructure licences in Hamburg, Munich, and Berlin.

The concentration of large financial firms in Paris makes this city an attractive prospect for COLT. In July 1996, COLT was granted a licence for this city and its nearby business district, La Defense. This licence enables COLT to provide switched and non-switched services within closed user groups. Later COLT was awarded a complete public network operator licence, including permission to offer voice telephony.

COLT's future plans include establishment in Spain, Italy, Belgium, the Netherlands, and Switzerland. This means that COLT will expand in Europe and add more geographical market segments, although the company continues to penetrate the chosen company customers in the relevant countries (Figure 10.7).

The divergence/adaptation cluster

Operators with the most market experience, no matter in which country, follow a resource-demanding strategy of divergence combined with a high degree of customization. Mercury (later a part of Cable & Wireless Communications) and Tele2 of Netcom Systems were the earliest challengers of the prior monopoly operators in the UK and Sweden: BT and Telia, respectively.

At the time of its entry into the market (1984), Mercury chose a twofold strategy. First, it would offer customers a combination of competitive prices and quality service, and it would be an innovative, proactive company. Second, it would enter a diverse range of markets. Although this policy spread resources over a wider area, it was deemed necessary to prevent BT from diverting its energy into Mercury's only chosen market and thus stifling growth.

However, the market and the competitive situation changed. Deregulation in the UK let in a wave of new entrants whose main edge was competitive pricing, as judged by Mercury. Consequently, there was constant price undercutting, and basic telephony prices fell. This meant that it was no longer viable for Mercury to try to compete primarily on price. Another significant change was that customer expectations increased greatly.

Thus, an internal evaluation of market policies and the recognition of a changing market environment caused Mercury to alter its business strategy. On the residential market, Mercury concentrated on customers in the target market whose bills were largely for long-distance and international calls. Regarding company customers, Mercury moved from trying to diversify itself across the market to focusing on particular niches. Mercury's customer service became based on responsiveness to customer requirements.

Tele2 started to offer data network services in Sweden in 1990 and addressed these to companies. Later, this operator also offered leased lines for national and international telephony. Broadscale marketing of public telephony services was initiated in 1993, when companies and private consumers were offered cheap international calls and, later, also national calls.

Tele2 views itself as an alternative to Telia and, accordingly, offers a broad range of telecommunications and data communications services, ranging from standardized telephony for small companies and private consumers to adapted communications solutions for large companies. Its intention is to keep its position as Sweden's largest privately owned telecommunications operator, and as a data communications actor.

Figure 10.7 illustrates the paths which Mercury and Tele2 have taken in their efforts to challenge the market dominants in the UK and Sweden.

National licences valid for, for example, mobile telephony include a requirement for national coverage. Therefore, the national network and infrastructure of a company such as Comviq GSM has to be built out in order to cover all regions of Sweden within a certain time limit. If a company wants to exploit its licence, the national infrastructure has to be built out, and this demands capital and puts a strain on profitability (Table 10.1).

Thus, once the licence conditions have been accepted, the investment process has to be initiated, and a company's freedom of action will be limited because of the increasing fixed costs. Accordingly, the strategy needs to be carefully formulated, and it is, more or less, a prerequisite that the intended strategy will be turned into a realized strategy.

COLT is an example of a company with a somewhat different strategy process, as its freedom of action is not as limited. COLT is

Table 10.1 Performance of companies operating telecommunications networks in the UK or Sweden (source: Annual Reports).

Performance measures[1]	*Mercury*	*COLT*	*Tele2*	*Comviq*
Turnover	1,654,8	9,2	72,1	83,7
Operating costs[2]	1,521,4	14,3	64,3	88,5
Operating profit (loss)	133,4	(5,1)	7,8	(4,8)
Profit margin, %	8	(55)	10.8	(5.7)

[1]1995 for Mercury, COLT and Tele2, and 1996 for Comviq.
[2]Operating expenses and depreciation.
GBP million (1 GBP=13 SEK).

gradually expanding in major European financial centers and, thus, is building out its infrastructure in a more step-by-step manner. In this process, each investment decision is not as risky as it is for Comviq GSM of Sweden. In each of the steps of the strategy development process, COLT generally has a number of alternatives to choose from; therefore, its strategy assumes an even more emergent character.

COLT and Comviq GSM are examples of operators trying to build out their own infrastructures in order to be able to attract customers and develop services. The expectation is that investments in networks will decrease and, as a result, that profitability will increase. Furthermore, as the desire is to restrict interconnection fees, the control of networks is meant to be a way of controlling costs in the long term.

In relative terms, Mercury and Tele2 in 1995 did not account for very high operating costs. To a large extent, this was due to relatively lower investments in infrastructure (Table 10.1), and it is appropriate to say that these operators have gained a foothold in the demanding strategic state of divergence/adaptation challenging the dominants in the UK and Sweden, respectively.

Anyhow, as I view it, a better understanding of history and of the emerging limits for freedom of action helps us to better appreciate the present and to more confidently predict the future. But the importance of intended and emerging strategy development generally differs from one company situation to another.

Competitive edge

Figure 10.8 illustrates the major competitive edges in the overview of the network operator industry in the UK and Sweden.

There is an obvious need for low prices as an initial means of penetrating telephony segments of the markets in both countries, even though telephony is only a part of a broader category of communications services. Price competition is particularly intense in international telephony and mobile telephony, characterized by standardized services that are difficult to differentiate and by numerous competitors.

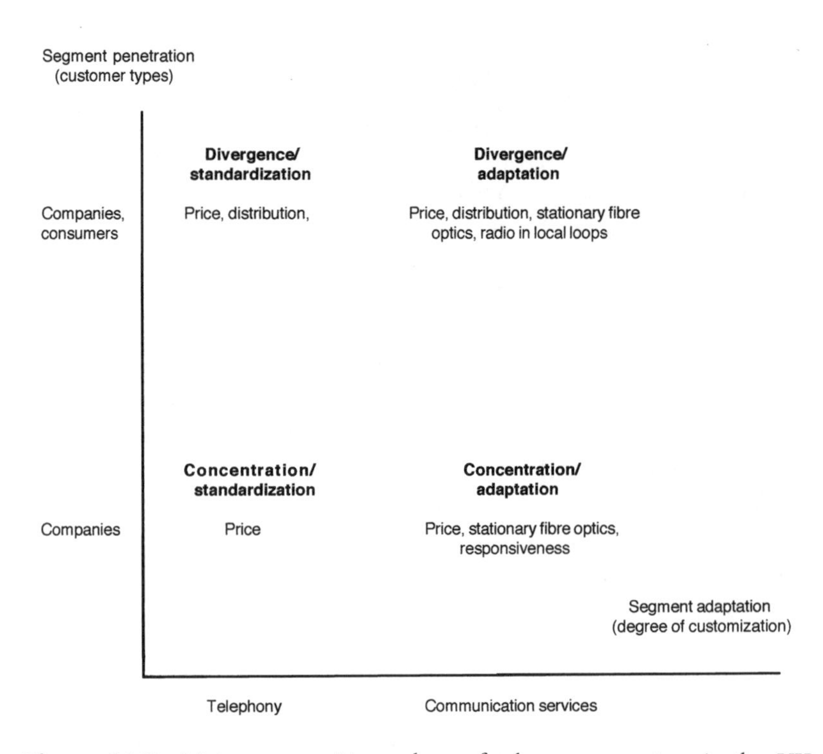

Figure 10.8 Major competitive edges of telecom operators in the UK and Swedish markets in 1995.

Besides low prices, efficient distribution systems are crucial for operators penetrating many market segments covering both companies and private consumers. This is particularly evident for mobile telephony operators, which address many market segments. For example, close cooperation with leading retail chains is regarded as the core of the marketing strategy of Comviq GSM. The company trains dealers' sales personnel and in other ways provides direct sales and marketing support. In the company market, a Comviq salesforce sells directly to customer corporations.

Comviq has chosen not to set up its own stores and instead works solely with independent dealers. These are divided into three groups: Independent retailers and buying chains; Mobile telephony chains; Radio and television outlets.

The first two groups have a strong hold on the company market, while radio and television outlets are dominants in the consumer market. In 1992, specialty stores accounted for nearly 80 percent of the total Swedish market, and radio and television outlets accounted for approximately 20 percent. These figures were almost completely reversed in 1996.

Transmission technology is constantly developing, and efforts to create alternative ways of reaching end-users are particularly evident in the strategic state of divergence/adaptation. For example, Tele2 of Sweden has signed an interconnection contract with Telia in order to be able to penetrate residentials and small companies. These customer types are normally not profitable enough to be directly linked to Tele2's stationary fibre optic network. However, the price levels stipulated in the Telia agreement makes it difficult for Tele2 to expand its national telephony operations. Thus, Tele2 has an explicit ambition of expanding its own network and of seeking alternative ways of accessing end-users. The use of radio technology in local loops is one example of a possible alternative in access networks.

Cable TV operators enjoy the freedom to offer both telephony and television services within licence areas in the UK. In London, Videotron's strategy of acquiring contiguous franchise areas facilitated construction of a single integrated fiber optic network. Thus, the broadband cable networks of Videotron made possible the provision of cable television and telecommunications services within the relevant licence areas. The customer groups consisted of residential telephony and cable TV customers and of business telecommunications customers.

Stationary fibre optics provides high capacity and reliable technology. COLT is an example of an operator situated in a strategic state of concentration/adaptation and that is competing with incumbent operators by emphasizing a competitive edge of high-quality services, primarily over its own fiber optic network. Responsiveness to the needs of the company's customers is an essential ingredient in COLT's marketing. This is manifested by such measures as analysis of the needs of the individual company, projecting activities, installation, and customer education activities. In essence, responsiveness here means an emphasis on the fulfillment of individual company needs.

We are now able to discuss generally the position of competitive edges in relation to the strategic states model. Thus, Figure 10.9 shows general positioning of competitive edges within the framework of the model.

The extreme states of A to D in Figure 10.9 demonstrate the need for different combinations of competitive edges. The relative standardization of products, as well as concentration on a limited number of market segments (state A), implies that competitive

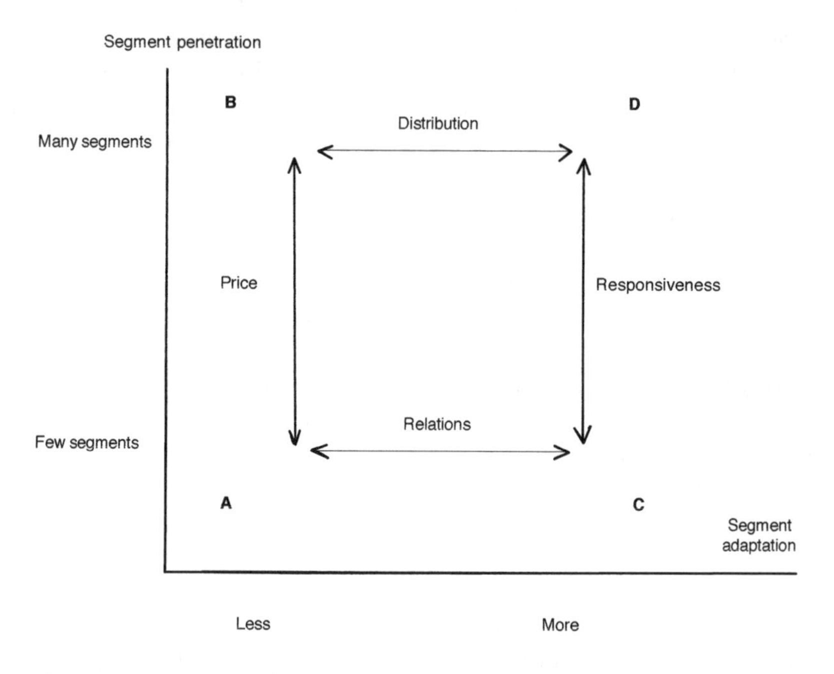

Figure 10.9 Crucial competitive edges within the strategic states model.

pricing is the main choice. The major reason for this is that customers do not perceive any important differences between competing products, except for price differences. Given price competition, high profit requires rationalization and conservative cost control and/or an increase of volume. Furthermore, it is crucial to try to create stable relationships with customers, as the risk of losing a single customer could cause serious effects, depending on the customer's importance. Such relationships normally make it possible to highlight competitive advantages other than price, which means that the company's vulnerability to price competition decreases.

The ability to form efficient distribution channels is crucial in the B state, where the company has a more divergent strategy but still offers standardized products. In fact, the ability to handle distribution issues related to the penetration of many market segments without any dominants will prove to be a major way of surviving and reaching high performance in price competition.

Responsiveness to customer needs resulting in customization of offerings makes it possible to avoid using price as the major competitive edge. However, the C state involves high risks from segment dependence and from the corresponding latent low bargaining power in business relations. Thus, there is a subsequent risk for limited freedom of action. A wise solution would be to try to incorporate a certain amount of standardization within the offered product range. In that case, products will be more easily available to other segments. However, too much standardization will result in losing the responsiveness advantage and instead will result in price competition.

Decentralization is the key to high performance in the D state. Extensive variation among customer requirements and many market segments makes it crucial to be able to get orders that are sufficiently similar to each other without losing the advantage of responsiveness. Otherwise, costs will probably be too high, as it will be difficult to exploit learning effects and to reduce variable costs. Thus, there is a need to define carefully the key core competencies that may be used by as many organizational units as possible. For instance, a capability of analyzing customer needs independent of which market segment is being treated may be regarded as a core competence in this state.

ELEVEN

Corporate Structure and Integration: Cable & Wireless vs. Netcom Systems

In the framework of strategic business units, formulation of corporate strategy usually applies to product and market issues relevant to an entire company group, while business strategy defines the choice of product, or service, and market of an individual business. Furthermore, Chakravarthy and Lorange (1991) talk about hierarchies of strategies, which are to be elaborated within the hierarchical strategic structure. On the corporate level, the main task, in theory, is to determine relationships among business families. This can be examined in several ways. Questions that may be asked include: Is there a satisfactory mix of mature and new businesses and between high-risk/high-return, and low-risk/low-return businesses? What is the nature of the overall pattern of funds flow? How do the businesses support each other in terms of human resources, know-how, and distinctive expertise? In particular, the organization at this level is finding out how the pieces all fit together.

I would like to add that corporate strategy formulation and implementation are strongly correlated with issues pertaining to corporate structure and integration. For example, if there is an explicit desire to exploit common experiences and bring consistent offerings to the market, forming of an appropriate organizational structure and implementing the accompanying integration measures are essential.

This chapter discusses key issues related to corporate structure and integration, which we are able to distinguish in the cases of Cable & Wireless of the UK and NetCom Systems of Sweden. Thus, the relevant processes are discussed in terms of observed causes of corporate restructuring and corporate integration together with examples of resulting measures. As the discussion on the cases indicates, restructuring and integration are important ingredients in the course of developing corporate strategy.

Cable & Wireless has combined organic growth of sole ventures and acquisitions, including mergers and alliances, while Netcom Systems has grown mainly through development of sole ventures. Both have experienced a need for corporate restructuring and integration but have applied somewhat different measures in the processes. Figure 11.1 gives examples of applied measures valid for some key empirically observed causes of corporate restructuring and integration.

Observed causes of corporate restructuring and integration	Examples of measures	
	Cable&Wireless	Netcom Systems
Corporate restructuring		
- Need to control businesses	Securing operating influence	Securing operating influence
- Need for an SBU structure	Merger, formation of business units	Formation of an operating group
Corporate integration		
- Need for cost reduction	Fewer duplications Key activities definition	Fewer duplications
- Need for coherent systems	Common network control and subscriber systems	Common network control systems
- Need for an explicit profile	Launch of a common brand	Launch of separate brands

Figure 11.1 Corporate restructuring and integration in Cable & Wireless of the UK and Netcom Systems of Sweden.

PROCESSES OF CORPORATE RESTRUCTURING

Referring to Figure 11.1, the following discussion comments on observations of corporate restructuring as regards the cases of Cable & Wireless and NetCom Systems.

Need to control businesses

The present CEO of Cable & Wireless was appointed in 1996, and he expressed a desire to control the businesses of the corporation, either by having controlling shares or by securing real operating influence:

> This way we can ensure all investment goes into businesses we control and influence on behalf of our shareholders. Where this is not possible we will exit and reinvest elsewhere. (CEO Richard H. Brown, Cable & Wireless plc, Annual Report, 1997)

In the autumn of 1996, Cable & Wireless announced that it had reached an agreement to create a large provider of integrated telecommunications, information, and entertainment services in the UK by forming Cable & Wireless Communications plc (CWC) from a merger of the operations of Mercury Communications Ltd, NynexCableComms Group plc, Nynex CableComms Group Inc., Bell Cablemedia plc, and Videotron Holdings plc. The new company would offer local, national, and international services in voice and data, together with multi-channel television, Internet services, and, in due course, interactive multimedia. "Cable & Wireless" constitutes the brand both for the group as a whole and for the UK company.

Considering the Swedish market, the dominating operator, Telia, is active in all segments of the market. However, competitive conditions vary from one segment to another. These variations are generally caused by different market growth, technical advances, or historic tariff structures. In 1993, when the Swedish market was further deregulated, Kinnevik opted to start individual companies in each segment and to adapt their strategies and allocate their resources on the basis of their potential. Tele2, Comviq GSM, and Kabelvision, thus, started to bring separate brand names to their respective markets.

At the same time, the telecommunications business area of Kinnevik became an operational unit, led by NetCom Systems AB, instead of a project organization as it had been before. With the purpose of consolidating and developing its position as the largest privately owned provider of telecom services in Scandinavia, a number of changes to the Group's structure were initiated in 1996. To start with, minority interests in subsidiaries were exchanged for shares in the parent company. As a result, NetCom Systems' Swedish telephony operations became wholly owned and its operating influence was secured:

> NetCom Systems has moved from being a "telecoms holding company" to being an operating group. The work of integrating operations in order to obtain synergies, both towards the market and within the organization, can now begin. (CEO Anders Björkman, NetCom Systems AB, Annual Report, 1996)

Need for an SBU structure

Both Cable & Wireless and NetCom Systems demonstrate, more or less explicitly, a need for implementation of structures relying on the concept of strategic business units (SBUs). For example, after the merger, CWC was initially divided into four business units, each penetrating residentials, small businesses, corporations, or international markets (primarily Ireland): Consumer; Business; Corporate; International&Partners (wholesale activities).

A managing director was appointed for each business unit. These managers were to be responsible for revenues and costs and for accompanying issues, such as product decisions, pricing, and operational marketing. Further, the business units would be supported by functions of customer operations and network and by staff in marketing, finance, communication, legal affairs, and so on.

In the group of NetCom Systems, individual operating companies have demonstrated an interest in similar customer groups. This makes possible the marketing of a broader range of services to customers in the existing customer base. In turn, this leads to a need for co-ordination of corporate activities pertaining to administration and sales. The market organization has, hence, been divided into a "company market" department and a "mass market" department. Thus, it is appropriate to say that this group implicitly needs an SBU structure.

There are, however, arguments against the usefulness of the SBU concept. For example, Prahalad and Hamel (1990) believe in a view of the company as a portfolio of competencies, besides portfolios of products and businesses. These authors point out that no single business in an SBU structure may feel responsible for maintaining a viable position in core products or be able to justify the investment required to build leadership in some core competencies. The people who embody this competence are typically seen as the sole property of the business in which they grew up. If core competencies are not recognized, individual SBUs will pursue only those innovation opportunities that are close at hand. Although I think that each company has to evaluate its own situation in this respect, core competencies and related activities should be assigned long-term perspectives and should generally be the responsibility of the corporate level.

If we consider the case of CWC, the question of whether the central staffs or whether the business units were to be responsible for certain key activities became a major issue in the restructuring process. For example, the central marketing staff was initially responsible for common branding (that is, for informing the market about the existence of the CWC brand and related issues), while the business units took care of customer relations. But how would this issue be treated in coming phases of the restructuring process? Other major questions of debate included the treatment of market segmentation. Which organizational level would be responsible for dividing potential and real customers into homogeneous categories?

PROCESSES OF CORPORATE INTEGRATION

A corporation may generally favor integration for reasons such as a need for more efficient utilization of existing resources and capabilities through economies of scale and for learning. By corporate integration, core competencies or other capabilities may be utilized over a broader base. Thus, integration is likely to be the desired means for exploiting the firm's existing resources, as well as for developing the firm's core competencies. In fact, Prahalad and Doz (1987) claim that, for example, intensive technology development encourages integration in that it facilitates quality control and co-ordination of new product introduction.

Observations of Cable & Wireless and NetCom Systems indicate that a need for cost reduction, a need for coherent systems, and a need for explicit profiles are relevant causes of corporate integration (Figure 11.1).

Need for cost reduction

As the desire for integration arises in response to pressures to utilize resources more efficiently, need for cost reduction is an essential cause of corporate integration. This citation underscores that a need for cost reduction may favor corporate integration:

> Cable & Wireless is a growth company in a growing market. The best measure of growth in a competitive marketplace is revenue and it will seek out creative and innovative ways to grow revenues. Management can also generate value by controlling costs and enhancing efficiency. Our intention is to widen the gap between costs and revenues year on year. One way to do that is by sharing learning experiences around the Group so that nothing is duplicated. (The Board of Directors, Cable & Wireless Communications plc, Annual Report, 1997)

Bringing together four companies in the UK gave access to around six million households and numerous businesses in a direct relationship not shared with any other provider. By running the combined traffic over one network, cutting out duplication, and standardizing services, without losing sight of the need to tailor services to local markets, the intention was to raise efficiency.

The total number of employees in the UK of the merged company, CWC, in the first year decreased from around 12,500 to 11,500. Essentially, duplications were taken away. Reduction of the number of call centers from 10 to three and a merger of five network control centers onto two sites are two examples of measures that resulted in cost reductions. Moreover, one intention is to integrate the four different subscriber management systems of CWC into a single system.

With the purpose of continuing rationalizations and reducing operating expenditures, key activities within CWC were identified, starting at the end of 1997. Senior management then set priorities among the primary activities as a basis for further decisions. In the review of resources, each organizational unit calculated resource demands pertaining to different service levels, such as the current level, enhanced levels, and a minimum level.

Need for coherent systems

A central question for a corporation competing in an emerging market is whether the various activities should be integrated into coherent systems or whether the single activities should be left to operate autonomously. From my point of view, systems generally contain continuous processes, such as planning or control, which essentially are a number of interrelated activities extended in time.

For example, CWC is implementing a common network control system and a single subscriber system. In Sweden, NetCom Systems is striving for the integration of its stationary, mobile and cable TV networks. This implies that fewer control stations will be needed in order to supervise network operations:

> NetCom's technical platform, which allows us to market seamlessly between the companies, is now in place. Examples of the services we will be offering include combined services with mobile/wired telephony and telephony/broadband connections to the Internet using cable TV networks. These services will allow us to break into areas where Telia has managed to keep its monopoly. (CEO Anders Björkman, NetCom Systems AB, Annual Report, 1996)

Need for an explicit profile

Whereas NetCom Systems has chosen to bring separate brands (Tele2, Comviq GSM, and Kabelvision) to the market, Cable & Wireless demonstrates a need for the marketing of an explicit profile and a common corporate brand.

A dominating percentage of the growth of Cable & Wireless previously came from businesses that did not carry the company name, and "Cable & Wireless" was introduced as the brand name for both the group as a whole and for the UK operation:

> One reason why we are regarded as the industry's best kept secret is that 80 per cent of our revenues have come from businesses which did not carry the Cable & Wireless name. To grow in a competitive world market place, we need to raise our profile; be one brand worldwide.
>
> Our name is truly international with no geographic restriction. It also describes what we do. So instead of facing the market with different logos, we're moving towards a common identity, consistent with being a single enterprise that thinks and acts as one. (CEO Richard H. Brown, Cable & Wireless plc, Annual Report, 1997)

In the UK, the four different brand names of the merged companies were simultaneously transferred to the single Cable & Wireless name. Simultaneously with the launch of the external brand, internal integration of processes, products, services, packages, and so on was started. This was combined with an extensive internal training program.

> Essentially, our profile is based on credibility. As regards services, we put forward flexibility and accessibility. Our intention is to be perceived as an inspiring company with personal relations to customers. (Andrew Law, Research&Analysis, CWC, 19 November 1997)

Hence, it is appropriate to say that the resulting explicit profile of Cable & Wireless in the UK comprises the interlinked components of the offering illustrated by Figure 11.2.

The company's capability of designing telephony and cable TV networks in order to be able to penetrate desired target segments and customers interested in end products such as telephony, data and TV services may be defined as the core competence. This competence clearly facilitates the penetration of a variety of market segments, and the quality of the networks (the core

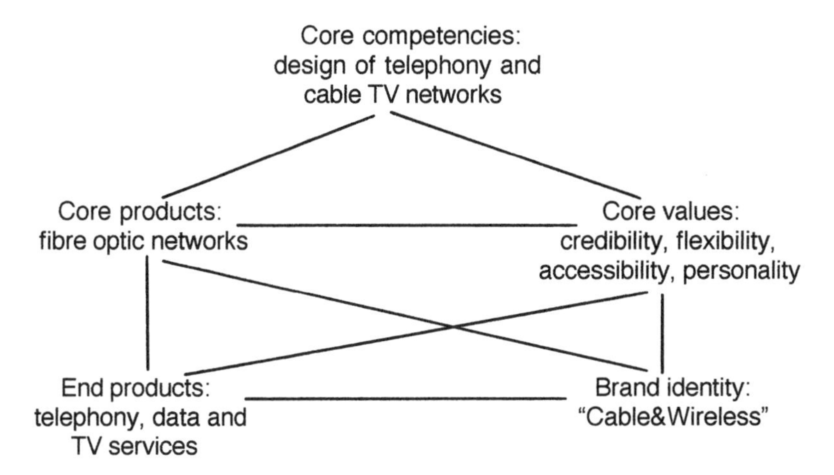

Figure 11.2 Components of a company offering: The example of Cable & Wireless Communications in the UK.

products) generally makes an important contribution to perceived customer benefits of the end products. Furthermore, part of the intricate issue of reaching end-users through the construction of local access networks and the inherent capital requirements show the difficulties to which any effort to imitate the core competence will give rise.

CWC mainly puts forward four core values: general credibility, flexibility and accessibility in the offered services, and a desire to establish personal relationships with customers. These values are continuously being integrated with the other components of the offerings, and the values define directions for product development, market communications, and customer interactions. Finally, the identity of the brand itself, "Cable & Wireless", is not geographically restricted, and it reflects the core competence of the corporation.

TWELVE

Concluding Remarks and Implications

The study of telecommunications operators in the UK and Sweden indicates strategy patterns that may occur in countries where deregulation has not developed very far. The other members of the European Union are examples of such countries. This study has not focused on every relevant factor that might have an impact on industry progress, however, it is likely that certain patterns will emerge in other countries. Based on a summary of the empirical findings of the present study, this chapter discusses the question of applicability of these findings to developments in other countries.

Besides telecommunications markets, there are several other markets that are being opened up. These can be defined by either technology or geographical specifications, or otherwise. For example, in many countries, electricity markets and transport markets relying on railways are emerging. Such markets previously have been unreachable for companies other than monopolies or other protected firms. This is also valid for geographical regions such as certain countries in Eastern Europe and Asia that become reachable by deregulation initiatives or by political developments in general. Hence, this chapter also reflects on generalizations of the findings to emerging markets other than telecommunications markets.

Finally, some advice is given, building on the present study of realized strategies, to management trying to cope with the difficult endeavor of formulating strategy in emerging markets in general.

APPLICABILITY OF THE FINDINGS TO OTHER EMERGING MARKETS

The present study generates a pattern for understanding how telecommunications operators have established themselves in the

UK and Sweden and for understanding which initial business strategies new entrants and previously established competitors follow. Furthermore, processes of corporate restructuring and integration in line with strategy development are exemplified, including observed causes of change and resulting measures. But to what extent do these empirical findings apply to emerging markets relevant to telecommunications operators in other countries, in other industries, or in other geographical regions?

Summary of patterns and processes detected in this study

In the market establishment model, which I have formed, perceptions of entry barriers imply varying assessments of business opportunities. These assessments influence the choice of entry strategy. Meanwhile, the process of strategy implementation may result in changing perceptions due to learning effects. Both perceptions and experiences gained during the implementation process are interrelated with comprehension of the need for strategy competence in order to become firmly established in the market in question. Of course, the character of relationships between perceptions of entry barriers, strategy competence, and the chosen entry strategy vary from one company case to another.

This study focuses on two important barriers to entering the telecommunications operators markets in the UK and Sweden: perception of deregulation and the capital needed for infrastructure investments. The processes of deregulation are perceived differently from one company to another, resulting in, for example, different opinions as regards the roles of the major actors, such as the regulators (Oftel and the National Post and Telecom Agency) and the market dominants (BT and Telia).

Besides the licence requirement, any company who wishes to enter the industry will face the need for capital, partly for the arrangement of infrastructure. No matter whether the operator tries to build its own network or whether it rents capacity, the capital demand is high. Networks based on fiber optic technology have, however, become a common substitute for networks of traditional copper cable, and this competition puts a strain on cost expansion. As direct connection to the new networks requires the penetration of volume customers in order to be profitable, large international companies are the main target group for direct fibre optic connections.

Local connections are frequently made possible through interconnections to existing local access networks based on copper cable. Together with the evolution of mobile telephony, however, radio technology is becoming a realistic alternative in local loops.

Strategy competence is the second component of the market establishment model. I argue that a certain company's strategy competence is due to the position of its businesses in relation to the original core business of the corporation (that is, "relatedness"). The underlying rationale is that high relatedness makes possible acquisition of related and relatively homogeneous knowledge of products, services, and markets. This means that efforts for establishing a business that may be considered a part of the original corporate core business hypothetically have the best chance of being successful. The opposite is true for a business unit that is less related to the core of the corporation.

The core business of a telecommunications operator would generally be the unit that is engaged in the construction of telephony networks. However, as core competencies develop and change over time, telecommunications operations of any corporation may be more or less related to the original corporate core business, which may be based on telephony networks design or on any other skill.

In the UK market, there are companies with telecommunications operations that do not belong to the original corporate core businesses. For example, electricity companies and cable TV companies based in the UK have found their way into the telecommunications industry using networks originally designed for other purposes. Cable TV companies exploit their fibre optic networks for local access and frequently provide telephony and related services in addition to what they have traditionally offered. This organic type of strategy development has not, according to this study, yet been seen on a broad scale in Sweden, although there are examples of co-operation between telecom operators and companies possessing other networks. Thus, in the Swedish market, almost every operator, including strategic alliances present in the market, belongs to a genuine telecommunications corporation with a high degree of relatedness among businesses.

Besides relatedness, strategy competence includes experience in the market. Familiarity with market conditions clearly strengthens the competence and the likelihood of finding a sustainable position in the market. Thus, increasing knowledge of products, services, and other market conditions gives extended market experience.

The patterns in market experience are basically of the same character in the UK and in Sweden. The dominants in the markets (BT and Telia) have their roots in the 19th century, when national monopolies were formed with the help of government initiatives. In the UK, however, the Kingston operator survived as a municipally owned company. The first real challengers, Mercury and Tele2, to the dominants in the UK and Sweden emerged in the 1980s and they belong to the group of early entrants. This is also valid for the first privately owned mobile telephony operators in both countries.

The late market entrants started to compete in the beginning of the 1990s in response to deregulation initiatives. This group comprises operators originating from the UK, Sweden or other countries; however, the original corporate core businesses vary somewhat from company to company.

As regards the final model component, entry strategy, my empirical findings show that the majority of operators entering the markets in the UK and Sweden initially prefer the organic development of sole ventures, whether the single operator is based in the local market or whether it is based in a foreign country. Although this is often a time-consuming entry mode, advantages such as the opportunity to build market experience in a controlled manner and to efficiently transfer technology know-how to the new ventures are obviously considered to be more important than gaining quick market shares through acquisitions or through forming alliances.

Patterns in initial business strategies are detected using the strategic states model, which I have developed. This model provides two dimensions along which clusters of competing operators are identified. In this study, the segment penetration dimension is measured by the customer types catered for (that is, whether the operators cater for company customers or for companies and consumers.) The dimension for the degree of adaptation of the offer takes two values: offering standardized telephony or offering broader communications services, including telephony.

Penetration of relatively few customer segments implies that careful building of relationships with customers is an essential measure in competition in a market. Conversely, having many customer types to be penetrated and reached underscores the importance of efficient distribution as a key competitive edge.

When it comes to the dimension of adaptation, standardization means that price competition may become severe and, consequently, competitive pricing is the real edge to be highlighted. A higher degree of adaptation ultimately requires responsiveness to customer needs in order to be successful.

The strategic state that is characterized by concentration on the company segment and by offering standardized telephony, was initially less occupied in both the UK and Sweden. Here, competitive pricing and an ability to create sustainable customer relationships are crucial for success.

Mobile telephony operators in both countries are situated in the state of divergence (penetration of companies and private consumers) and standardization. Frequently, mobile telephony operators co-operate with operators of stationary telephony in order to be able to offer more complete product ranges. In this strategic state, efficient distribution together with attractive prices are key competitive edges.

All of the UK and Swedish operators that are located in the state of concentration on companies and of offering broader customized communications services, including telephony, entered the market late. This indicates that adaptation to certain company customers has become a popular initial business strategy among new entrants, but this requires an ability to respond to customer needs and to skillfully build relationships in order for the strategy combination to be successful.

Operators with the most market experience, no matter in which country, follow a resource-demanding strategy of divergence combined with a high degree of customization. Here, the extensive variation among customer requirements makes it necessary to be able to get orders with the help of an efficient distribution system that are similar enough to each other that learning effects can be exploited without losing the advantage of highly developed responsiveness.

Further, corporate strategy formulation and implementation is strongly correlated with issues pertaining to corporate structure and integration. For example, if there is an explicit desire to exploit common experiences and to bring consistent offerings to the market, forming of an appropriate organizational structure and implementing the accompanying integration measures are essential.

Cable & Wireless of the UK and NetCom Systems of Sweden have expressed a desire for corporate restructuring and integration, but they have applied somewhat different measures in their efforts to get complete control over businesses, to form structures similar to those of strategic business units, to reduce costs, to build coherent management systems, and to create explicit corporate profiles.

One of the central issues that separates Cable & Wireless and NetCom Systems is whether central staffs or business units will be responsible for certain key activities, such as brand launching. Whereas NetCom Systems brings different brands to the market, Cable & Wireless has chosen a common corporate brand in order to facilitate its efforts to form a more explicit corporate profile. In my opinion, this expresses the core competencies (design of telephony and cable TV networks), core products (fiber optics networks), end products (telephony, data and TV services), core values (credibility, flexibility, accessibility, personality), and a brand identity ("Cable & Wireless").

Are the findings applicable to other emerging markets?

This study shows the robustness of both the market establishment model and the strategic states model. The model applications generate some findings that are of a general character, and the same operationalizations may be used in studies of other emerging markets.

When it comes to the market establishment model, however, entry barriers need to be relevant to the industry or region studied. Perceptions of deregulation and capital needed for infrastructure investment are, for example, relevant to telecommunications operators in all European Union countries but not necessarily to other industries or regions. Regardless of the choice of variables, entry barriers are generally perceived differently from one company to another and from one individual to another, partly because of the uncertainty inherent in environmental judgments.

Although I have not statistically tested the assumption that efforts to establish a business that is highly related to the original corporate core business are more likely to be successful than are efforts to establish a business that is less related to the core, the findings of this study support the hypothesis. Consequently, I assume that high relatedness among businesses is a crucial strategy competence factor in other emerging markets. This also applies to market experience, including knowledge of relevant products, services, target customers, and other key market conditions.

Operators entering the UK or Swedish markets most commonly choose the sole venture mode of organic development, but this does not necessarily mean that this pattern will be repeated in other countries or industries. Furthermore, as entry mode is just one part of a more comprehensive entry strategy, the findings of this study cannot be generalized to broader entry strategy contexts. We need more empirical information on other aspects of entry strategy, such as company goals, resource allocations, performance monitoring, and time schedules.

Considering the strategic states model, the empirical patterns detected depend, to some extent, on which measurements and which variable values one chooses in applications. As regards the industry of telecommunications operators, it seems that at least one finding may be valid for countries other than the UK and Sweden. That is, adaptation to well-defined company customers is a common immediate business strategy of new entrants. If a newcomer gets a foothold in this strategic state, it then may challenge previously established competitors in other states while defending its initial position. One way to get desired company customers is to offer direct connections to fiber optic networks and broad communications services, including telephony. In this strategic state, it is necessary to be able to carefully observe customer needs and to build sustainable relationships with important customers.

As we know from previous research, structure and integration aspects have to be reflected upon in the course of studying corporate strategy in general, independent of the industry or market character. Some causes of corporate restructuring and integration have been observed through case analyses in this study (the need to control businesses and to introduce a business unit structure, and the need for cost reduction, coherent systems, and an explicit corporate profile). I would like to add to the general discussions on corporate integration by saying that an explicit corporate profile ideally should reflect the interlinked components of core competencies, core products, end products, core values, and brand identity.

IMPLICATIONS FOR MANAGEMENT

Although we have to remember that changing conditions make the formulation of strategy in emerging markets a difficult endeavor, it is possible to put forward some suggestions based on this study. No matter which emerging market we are focusing on, the market establishment model and the strategic states model may be valuable in strategy formulation.

A company evaluating an establishment in an emerging market needs to carefully assess relevant barriers to exploiting business opportunities. On the other hand, if the company is already established in the market, the existence of barriers is crucial to the competitive situation in the market, and the company in question indeed has to be aware of what, if anything, protects the industry from extended competition.

Strategy competence is decisive in trying to exploit business opportunities. Here it is of great interest to make a judgment on the importance of relatedness among corporate businesses and pertinent market experience. If sufficient competence is not available, a decision has to be made on how relevant competence will be acquired.

Simultaneously with assessing entry barriers and the strategy competence required to become firmly established in the emerging market, an appropriate initial entry strategy needs to be chosen. This decision involves considerations such as inherent time horizons, the choice of entry mode, and performance monitoring. If the company is not a new entrant, it perhaps needs to reconsider its initial mode of entry and to change its attitude towards sole ventures, strategic alliances, or acquisitions in order to keep or alter its market position.

As regards the choice of current business strategy in the emerging market, a market segmentation procedure will indicate which parts of the market are best suited for penetration. Furthermore, information on real customer needs and requirements is the basis for choosing to which degree the offerings should be adapted. After that, the company will be capable of choosing a pure business strategy, such as concentration, divergence, standardization, or adaptation, or any combination of these. However, no matter which strategy or strategy combination the company chooses, the choice brings an emphasis on certain competitive edges.

It is not possible to isolate strategy choices from other areas in company development. Each strategy chosen affects, for example, organizational structures and the treatment of company integration. In the process of restructuring and integration, it is important to be aware of what the real needs are before measures are decided. These may, though, concern issues related to operational control, structural formation, cost reduction, administrative systems, branding or other crucial issues.

References

BOOKS AND ARTICLES

Albaum, G., Strandskov, J., Duerr, E. and Dowd, L. (1994) *International Marketing and Export Management*, Wokingham, England: Addison-Wesley.

Andrews, K. (1965) *Business Policy: Texts and Cases*, Homewood, IL: Irwin.

Andrews, K. (1987) *The Concept of Corporate Strategy*, Homewood, IL: Irwin.

Ansoff, I. (1965) *Corporate Strategy: An Analytical Approach to Business Policy for Growth and Expansion*, New York: McGraw-Hill.

Ansoff, I. (1991) "Critique of Henry Mintzberg's 'The design school': reconsidering the basic premises of strategic management", *Strategic Management Journal*, 12, pp. 449–461.

Bain, J. (1956) *Barriers to New Competition*, Cambridge, MA: Harvard University Press.

Barnett, W. and Burgelman, R. (1996) "Evolutionary perspectives on strategy", *Strategic Management Journal*, 17, Summer Special Issue, pp. 5–19.

Beesley, M. and Laidlaw, B. (1989) "The future of telecommunications: an assessment of the role of competition in UK policy", *Research Monograph Series No. 42*, London: Institute of Economic Affairs.

Bergendahl-Gerholm, M. and Hultkrantz, L. (1996) "Next step in telecom politics" (in Swedish), *Ds 1996: 29*, Stockholm: Ministry of Finance, The Expert Group for Studies in Public Economy,

Bohlin, E. and Granstrand, O. (eds.) (1994) *The Race to European Eminence: Who are the Coming Tele-service Multinationals*?, Amsterdam: North-Holland.

Bourgeois III, L. (1985) "Strategic goals, perceived uncertainty and economic performance in volatile environments", *Academy of Management Journal*, 28, pp. 548–573.

Burton, J. (1997) "The competitive order or ordered competition?", *Public Administration*, 75, pp. 57–89.

Buzzell, R., Gale, B. and Sultan, R. (1975) "Market share: a key to profitability", *Harvard Business Review*, No. 1, pp. 97–106.

Carleheden, S. (1999) *Strategies of the Telephony Monopolies* (diss., in Swedish), Lund, Sweden: Lund University Press.

Cave, M. and Sharma, Y. (1994) "Foreign entry and competition for local telecommunications services in the UK after the duopoly review", in Bohlin, E. and Granstrand, O. (eds), *The Race to European Eminence: Who are the Coming Tele-service Multinationals?*, Amsterdam: North-Holland.

Chakravarthy, B. and Lorange, P. (1991) *Managing the Strategy Process*, Englewood Cliffs, NJ: Prentice-Hall International.

Chandler, A. (1962) *Strategy and Structure: Chapters in the History of American Enterprise*, Cambridge, MA: MIT Press.

Chang, S. (1996) "An evolutionary perspective on diversification and corporate restructuring: entry, exit and economic performance during 1981–89", *Strategic Management Journal*, 17, pp. 587–611.

Dang-N'guyen, G. and Phan, D. (1994) "Competition in the British Telephony Market", in Bohlin, E. and Granstrand, O. (eds), *The Race to European Eminence: Who are the coming Tele-service Multinationals?*, Amsterdam: North-Holland.

Drucker, P. (1954) *The Practice of Management*, New York: Harper and Row.

Ellis, J. and Williams, D. (1995) *International Business Strategy*, London: Pitman Publishing.

Faulkner, D. and McGee, J. (1997) "Success and failure of international strategic alliances: evidence from in-depth case studies", in Thomas, H., O'Neal, D. and Ghertman, M. (eds), *Strategy, Structure and Style*, Chichester, England: Wiley.

Fayerweather, J. (1969) *International Business Management: A Conceptual Framework*, New York: McGraw-Hill.

Fransman, M. (1997) "Towards a new agenda for Japanese telecommunications", *Telecommunications Policy*, 21, pp. 185–194.

Galbraith, C. and Schendel, D. (1983) "An empirical analysis of strategy types", *Strategic Management Journal* , 4, pp. 153–173.

Ginsberg, A. and Venkatraman, N. (1985) "Contingency perspectives of organizational strategy: a critical review of the empirical research", *Academy of Management Review*, 10, pp. 421–434.

Granstrand, O. and Johansson, O. (1994) "Internationalization of the Swedish telecom services market", in Bohlin, E. and Granstrand, O. (eds), *The Race to European Eminence: Who are the Coming Tele-service Multinationals?*, Amsterdam: North-Holland.

Grant, R. (1991) "The resource-based theory of competitive advantage: implications for strategy formulation", *California Management Review*, 33, pp. 114–135.

Hall, W. (1978) "SBUs: Hot, new topic in the management of diversification", *Business Horizons*, February, pp. 17–25.

Hambrick, D. and Lei, D. (1985) "Toward an empirical prioritization of contingency variables for business strategy", *Academy of Management Journal*, 28, pp. 763–788.

Hamel, G. (1991) "Competition for competence and interpartner learning within international strategic alliances", *Strategic Management Journal*, 12, pp. 83–103.

Hamel, G. and Prahalad, C. (1993) "Strategy as stretch and leverage", *Harvard Business Review*, March/April, pp. 75–84.

Haspeslagh, P. (1982) "Portfolio planning: uses and limits", *Harvard Business Review*, January/February, pp. 58–73.

Hatten, K., Schendel, D. and Cooper, A. (1978) "A strategic model for the U.S. brewing industry: 1952–1971", *Academy of Management Journal*, 21, pp. 592–610.

Hofer, C. (1975) "Toward a contingency theory of business strategy", *Academy of Management Journal*, 18, pp. 784–810.

Hofer, C. and Schendel, D. (1978) *Strategy Formulation: Analytical Concepts*, New York: West Publishing Company.

Johansson, O. (1994) "Internationalization and diversification of technology-based services: a comparative analysis of 25 telecommunication service corporations", in Bohlin, E. and Granstrand, O. (eds), *The Race to European Eminence: Who are the Coming Tele-service Multinationals?*, Amsterdam: North-Holland.

Karlsson, M. (1998) "The liberalisation of telecommunications in Sweden: technology and regime change from the 1960s to 1993", *Linköping Studies in Arts and Science No. 172*, Linköping, Sweden: Linköping University.

Kissinger, H. (1977) *American Foreign Policy*, New York: W. W. Norton.

Kramer, R. and NiShuilleabhain, A. (1994) "Cable&Wireless: services, investments prospects", in Bohlin, E. and Granstrand, O. (eds), *The Race to European Eminence: Who are the Coming Tele-service Multinationals?*, Amsterdam: North-Holland.

Lorange, P. (1980) *Corporate Planning: An Executive Viewpoint*, Englewood Cliffs, NJ: Prentice Hall.

Lorange, P. (1984) "Organizational structure and management processes: implications for effective strategic management", in

Guth, W. (ed.), *Handbook on Strategic Management*, New York: Van Nostrand.

Lorange, P. and Roos, J. (1993) *Strategic Alliances: Formation, Implementation and Evolution*, Oxford, England: Blackwell

Mansell, R. (1993) *The New Telecommunications: a Political Economy of Network Evolution*, London: Sage Publications.

Marsh, S. (1998) "Creating barriers for foreign competitors: a study of the impact of anti-dumping actions on the performance of U.S. firms", *Strategic Management Journal*, 19, pp. 25–37.

Miller, D. (1988) "Relating Porter's business strategies to environment and structure: analysis and performance implications", *Academy of Management Journal*, 31, pp. 280–308.

Milliken, F. (1987) "Three types of perceived uncertainty about the environment: state, effect and response uncertainty", *Academy of Management Review*, 12, pp. 133–143.

Mintzberg, H. (1990a) "Strategy formation: schools of thought", in Frederickson, J. (ed.), *Perspectives on Strategic Management*, New York: Harper and Row.

Mintzberg, H. (1990b) "The design school: reconsidering the basic premises of strategic management", *Strategic Management Journal*, 11, pp. 171–195.

Mintzberg, H. and Waters, J. (1985) "Of strategies: deliberate and emergent", *Strategic Management Journal*, 6, pp. 257–272.

Mölleryd, B. (1997) "The building of a global industry: the importance of entrepreneurship for Swedish mobile telephony", *Via Teldok*, 28.

Noda, T. and Bower, J. (1996) "Strategy making as iterated processes of resource allocation", *Strategic Management Journal*, 17, pp. 159–192.

O'Reilly, D. (1995) "Classical competitive strategy in newly deregulated industries: does it apply?" in Hussey, D. (ed.), *Rethinking Strategic Management*, Chichester, England: Wiley.

Parasiris, E. (1995) "Change processes within the telecommunications services market" (in Swedish), *Research Report No 1995: 14*, Stockholm: School of Business.

Pehrsson, A. (1985) *Strategic Planning and Environmental Judgements: the Performance in S.B.U. Organized Industrial Groups* (diss.), Linköping, Sweden: Linköping University.

Pehrsson, A. (1990) "Strategic groups in international competition", *Scandinavian Journal of Management*, 6, pp. 109–124.

Pehrsson, A. (1993) "A contingency perspective of strategy choice problems: experiences of Swedish companies in Germany", *Journal of Strategic Change*, 2, pp. 89–101.

Pehrsson, A. (1995a) "Strategic states and performance: Swedish companies in Germany", *Journal of Strategic Change*, 4, pp. 229–237.

Pehrsson, A. (1995b) "International product strategies: an exploratory study", *Scandinavian Journal of Management*, 11, pp. 237–249.

Pehrsson, A. (1996) "International strategies in telecommunications: model and applications", *Routledge Research in Organizational Behaviour and Strategy No. 2*, London: Routledge.

Pehrsson, A. (2000) "Strategy competence: a key profitability driver", *Strategic Change*, 9, March/April, pp. 89–102

Penrose, E. (1980) *The Theory of the Growth of the Firm*, Oxford: Blackwell Publishers.

Perlmutter, H. (1969) "The tortuous evolution of the multinational corporation", *Columbia Journal of World Business*, pp. 9–18.

Porter, M. (1980) *Competitive Strategy*, New York: The Free Press.

Porter, M. (1985) *Competitive Advantage*, New York: The Free Press.

Porter, M. (1990) *The Competitive Advantage of Nations*, London: MacMillan.

Prahalad, C. and Hamel, G. (1990) "The core competence of the corporation", *Harvard Business Review*, May/June, pp. 79–91.

Prahalad, C. and Doz, Y. (1987) *The Multinational Mission*, New York: The Free Press.

Prosser, T. (1997) *Law and the Regulators*, Oxford: Oxford University Press.

Root, F. (1987) *Entry Strategies for International Markets*, Lexington, MA: Lexington Books.

Rumelt, R. (1974) *Strategy, Structure and Economic Performance*, Cambridge, MA: Harvard University Press.

Rumelt, R. (1982) "Diversification strategy and profitability", *Strategic Management Journal*, 4, pp. 359–369.

Rutihinda, C. (1996) *Resource-Based Internationalization: Entry Strategies of Swedish Firms into the Emerging Markets of Eastern Europe*, (diss.), Stockholm: Stockholm University.

Samiee, S. and Roth, K. (1992) "The influence of global marketing standardization on performance", *Journal of Marketing*, April, pp. 1–17.

Scott, C. (1996) (ed.) *International Regulatory Competition and Coordination*, Oxford: Oxford University Press.

Shepherd, W. (1997) "Effective competition in telecommunications, railroads and electricity", *Antitrust Bulletin*, 42, pp. 151–176.
Simon, H. (1962) "The architecture of complexity", *Proceedings of The American Philosophical Society*, 6, pp. 467–482.
Spiller, P. and Cardilli, C. (1997) "The frontier of telecommunications deregulation: small countries leading the pack", *Journal of Economic Perspectives*, 11, pp. 127–139.
Sutcliffe, K. and Zaheer, A. (1998) "Uncertainty in the transaction environment: an empirical test", *Strategic Management Journal*, 19, pp. 1–23.
Sölvell, Ö, Zander, I. and Porter, M. (1991) *Advantage Sweden*, Stockholm: Norstedts.
Trebing, H. and Estabrooks, M. (1995) "The globalization of telecommunications: a study in the struggle to control markets and technology", *Journal of Economic Issues*, 29, pp. 535–545.
Urde, M. (1997) *Brand Orientation: Development of Brands as Strategic Resources and Protection Against Trademark Degeneration* (diss., in Swedish), Lund: Lund University Press.
Valletti, T. and Cave, M. (1998) "Competition in UK mobile communications", *Telecommunications Policy*, 22, pp. 109–132.
Waverman, L. and Sirel, E. (1997) "European telecommunications markets on the verge of full liberalization", *Journal of Economic Perspectives*, 11, pp. 113–127.
Williams, H. and Taylor, J. (1994) "Competencies and diversification: the strategic management of BT", in Bohlin, E. and Granstrand, O. (eds), *The Race to European Eminence: Who are the Coming Tele-service Multinationals?*, Amsterdam: North-Holland.
Yip, G. (1982) "Diversification entry: internal development versus acquisition", *Strategic Management Journal* , 3, pp. 331–345.

SOURCES FOR EMPIRICAL INFORMATION

Annual Reports (1995) of Atlantic Telecommunications Ltd, Banverket, British Telecommunications plc, Dotcom Data&Telecommunications AB, France Telecom (Global One), Kingston Communications (Hull) Ltd, Mercury Communications Ltd, Mercury Personal Communications Ltd, MFS Communications, Nordiska Tele8 AB, Orange Personal Communications Services, Racal Network Services Ltd, Scottish

Hydro-Electric plc, Scottish Power Telecommunications Ltd, Telecom Finland AB, Telenordia AB, Telstra UK Ltd, Videotron Ltd, Vodafone Ltd, Worldcom International Inc.

Annual Reports (1995 and 1996) of COLT Telecommunications Group plc, NordicTel Holdings AB, Telia AB.

Annual Reports (1995–1997) of Cable&Wireless Communications plc.

Annual Reports (1992–1996) of NetCom Systems AB.

COLT Telecommunications Newsletter, Issue 4, Autumn 1995.

COLT Telecommunications Newsletter, Issue 8, Autumn 1997.

Report and Financial Statements, Videotron Holdings plc, 31 August 1996.

"The state applies to a global telecom giant" (in Swedish) (1996a) *Svenska Dagbladet*, 22 February.

"Tele8 has difficulties to get started" (in Swedish) (1996b) *Svenska Dagbladet*, 27 July.

"Low prices a weapon in global telecom competition" (in Swedish) (1996c) *Svenska Dagbladet*, 14 October.

"The best alternative" (in Swedish) (1996d) *Svenska Dagbladet*, 27 November.

"Competition in UK Telephony" (1997), *CIT Publications*, February.

Utility Week (1997) January 17.

A Survey of Competition and Development on the Swedish Telecoms Market in 1996 (in Swedish) (1997), Stockholm: The National Post and Telecom Agency.

Teldok Info (in Swedish) (1994) Stockholm: Telia AB, May.

Oftel Statement: Effective Competition, Framework for Action (1995), London: Office of Telecommunications, July.

Oftel Annual Report 1996 (1997), London: Office of Telecommunications, March.

Personal interviews:

— Andrew Law, Head of Research&Analysis, Cable&Wireless Communications, London, 19 November 1997 and 27 March 1998.

— Michael Bryan-Brown, Regulatory Affairs, COLT Telecommunications, London, 1 December 1997.

— Pelle Hjortblad, Market Director, Tele2 AB, Stockholm, 29 October 1997.

Index